Design Structure Matrix Methods and Applications

T0333714

Flexibility in Engineering Design, by Richard de Neufville and Stefan Scholtes, 2011

Engineering a Safer World, by Nancy G. Leveson, 2011

Engineering Systems, by Olivier L. de Weck, Daniel Roos, and Christopher L. Magee, 2011

Design Structure Matrix Methods and Applications, by Steven D. Eppinger and Tyson R. Browning, 2012

Design Structure Matrix Methods and Applications

Steven D. Eppinger and Tyson R. Browning

The MIT Press
Cambridge, Massachusetts
London, England

First MIT Press paperback edition, 2016

This book was set in Syntax and Times Roman by Toppan Best-set Premedia Limited. Printed and bound in the United States of America.

Library of Congress Cataloging-in-Publication Data

Eppinger, Steven D.
Design structure matrix methods and applications / Steven D. Eppinger and Tyson R. Browning.
    p.   cm. — (Engineering systems)
Includes bibliographical references and index.
ISBN 978-0-262-01752-7 (hardcover : alk. paper) — 978-0-262-52888-7 (pb)
1. Product design. 2. Systems engineering. 3. Flexible manufacturing systems. I. Browning, Tyson R., 1971–
II. Title.
TS171.E67   2012
670.42′7 — dc23
2011039033

To Don Steward, grandfather of DSM

Contents

1   Introduction to Design Structure Matrix Methods

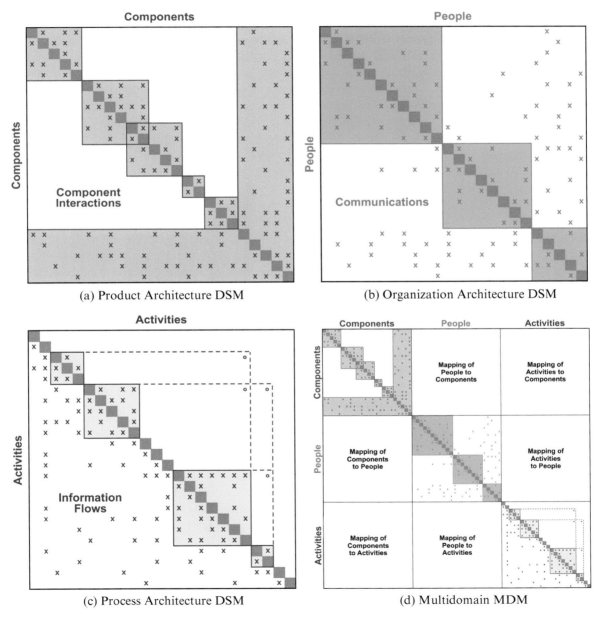

(a) Product Architecture DSM

(b) Organization Architecture DSM

(c) Process Architecture DSM

(d) Multidomain MDM

Figure 1.1
The four primary types of DSM models discussed in this book.

The Complex World of Systems

Our world is growing more complex every day. As we discover more and more about nature at both the micro and macro levels, we accumulate an exponentially increasing amount of information. From subatomic particles to galaxies farther away, we observe and record more and more data. This information empowers us to design and build ever more complex artificial systems. From aircraft, automobiles, computers, electronics systems, ships, machine tools, and buildings to sociotechnical systems, we continue to improve on ways to get a variety of people, materials, and instructions to work together to provide capabilities that they could not achieve separately. Learning from and about these systems provides even more information. As information systems such as the Internet have enabled so much information to disseminate, more and more people are empowered to contribute to this process of information generation.

Today, we are already overwhelmed with more information than we can digest, so we turn to search engines and filters to help us access the information we want—or think we want. But as the amount of information continues to grow, and as any one person or group's knowledge and information-processing capabilities are limited, when it comes to getting the right information to the right place at the right time, the chances of error are growing. We must continue to develop techniques for mastering the vast amount of information required to understand, design, and improve systems. Because no one person knows enough to design today's complex systems, useful techniques for managing information must draw out the knowledge from individuals and cast it in a way that enables a transdisciplinary group to review and critique it. This book is about one such technique that has been used to help people better design, develop, and manage complex engineered systems such as the ones pictured in figure 1.2. This technique is known as the design structure matrix (DSM).

What Is the DSM?

The DSM is a network modeling tool used to represent the elements comprising a system and their interactions, thereby highlighting the system's architecture (or designed structure). DSM is particularly well suited to applications in the development of complex, engineered systems and has to date primarily been used in the area of engineering management. On the horizon, however, is a much broader range of DSM applications addressing complex issues in health care management, financial systems, public policy, natural sciences, and social systems.

The DSM is represented as a square $N \times N$ matrix, mapping the interactions among the set of N system elements. A highly flexible tool, DSM has been used to model many types of systems. Depending on the type of system being modeled, DSM can represent various types of architectures. For example, to model a product's architecture, the DSM

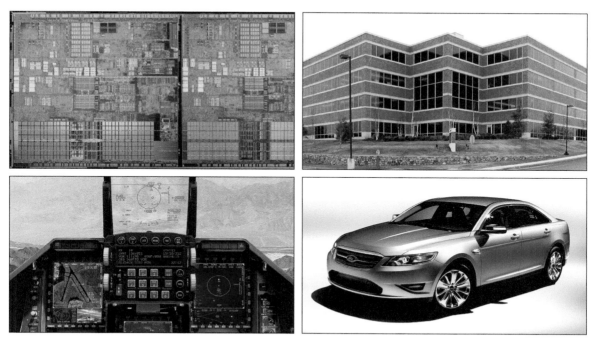

Figure 1.2
Some of the complex systems modeled using DSM: electronics systems (see examples 3.3, 3.9, 7.2, 7.8, 7.9, 9.10), buildings (see examples 3.8, 5.5, 7.1, 7.4, 7.7), aircraft (see examples 3.2, 3.3, 5.2, 5.3, 7.6, 7.10, 7.11, 9.2, 9.12), and automobiles (see examples 3.1, 5.1, 7.12, 9.1, 9.4, 9.6, 9.11).

elements would be the components of the product, and the interactions would be the interfaces between the components (figure 1.1a). To model an organization's architecture, the DSM elements would be the people or teams in the organization, and the interactions could be communications between the people (figure 1.1b). To model a process architecture, the DSM elements would be the activities in the process, and the interactions would be the flows of information and/or materials between them (figure 1.1c). DSM models of different types of architectures can even be combined to represent how the different system domains are related within a larger system (figure 1.1d). Thus, the DSM is a generic tool for modeling any type of system architecture. In this book, we discuss how DSM has been used in all of these domains and more.

Compared with other network modeling methods, the primary benefit of DSM is the graphical nature of the matrix display format. The matrix provides a highly compact, easily scalable, and intuitively readable representation of a system architecture. Figure 1.3a shows a simple DSM model of a system with eight elements, along with its equivalent directed graph (digraph) representation in figure 1.3b. When one is first introduced to the

DSM, many find it easy to think of the cells along the diagonal of the matrix as representing the system elements—analogous to the nodes in the digraph model. To keep the matrix diagram compact, the full names of the elements are often listed to the left of the rows (and sometimes also above the columns) rather than in the diagonal cells. It is also easy to think of each diagonal cell as potentially having inputs entering from its left and right sides and outputs leaving from above and below. The sources and destinations of these input and output interactions are identified by marks in the off-diagonal cells—analogous to the directional arcs in the digraph model. Examining any row in the matrix reveals all of the inputs to the element in that row (which are outputs of other elements). Looking down any column of the matrix shows all of the outputs from the element in that column (which become inputs to other elements).

In the simple DSM example shown in figure 1.3a, the eight system elements are labeled A through H, and we have labeled both the rows and columns A through H accordingly. Reading across row D, for example, we see that element D has inputs from elements A, B, and F, represented by the X marks in row D, columns A, B, and F. Reading down column F, we see that element F has outputs going to elements B and D. Thus, the mark in the off-diagonal cell [D, F] represents an interaction that is both an input and an output depending on whether one takes the perspective of its provider (column F) or its receiver (row D).

It is important to note that many DSM resources use the opposite convention, the transpose of the matrix, with an element's inputs shown in its column and its outputs shown in its row. The two conventions convey the same information, and both are widely used because of the diverse roots of matrix-based tools for modeling systems (which is the topic

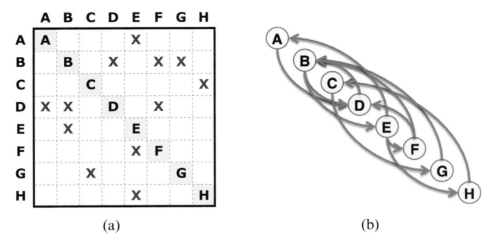

(a)   (b)

Figure 1.3
The binary DSM (a) with inputs in rows (IR) and its equivalent in digraph form (b).

of the next section). Here we begin by adopting the original DSM convention with inputs in rows (IR) and outputs in columns. Later we also present examples using the opposite inputs in columns (IC) convention (which stems from N^2 charts and IDEF0 diagrams).

The simple DSM in figure 1.3a is called a *binary DSM* because the off-diagonal marks indicate merely the presence or absence of an interaction. The binary DSM representation can be extended in a great variety of ways by including further attributes of the interactions, such as the number of interactions and/or the importance, impact, or strength of each—which might be represented by using one or more numerical values, symbols, shadings, or colors instead of just the binary marks in each of the off-diagonal cells. This extended form of DSM is called a *numerical DSM*. Figure 1.4 shows two examples. Additional attributes of the elements themselves may also be included by adding more columns to the left of the square matrix to describe, for example, the type, owner, or status of each element. (Additional attributes of the interactions, such as their names, requirements, etc. are usually kept in separate repositories but may be linked to the DSM cells by numerical identification numbers or indices.)

DSM models can be partitioned or rearranged using a variety of analytical methods, the most common of which are clustering and sequencing, as shown in figures 1.5a and b, respectively. Clustering analysis applies primarily to the kinds of interaction networks found in product and organization architecture DSM models, where interaction marks are largely symmetric about the diagonal, as described in chapters 2–5. Sequencing analysis applies primarily to the kinds of directional or temporal interaction networks found in process DSM models, as described in chapters 6–7.

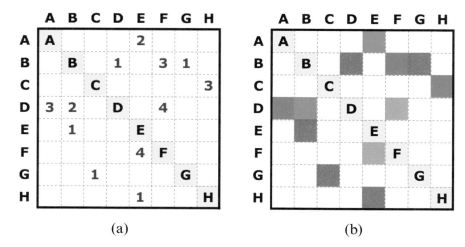

(a)   (b)

Figure 1.4
The numerical DSM representation using values (a) or colors (b) to represent strength or type of interactions.

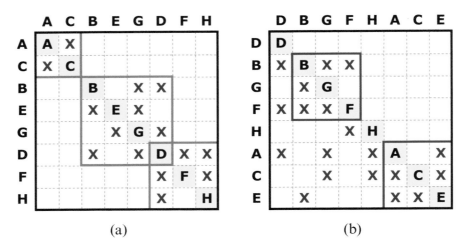

Figure 1.5
DSM partitioning analysis commonly entails clustering (a) or sequencing (b) based on the interactions contained in the matrix. (To illustrate clustering and sequencing, the interaction data in these two matrices are not the same and are different than those in figures 1.3 and 1.4.)

Matrix-Based Tools for Modeling System Architectures

The term DSM has its basis in using a matrix to model the design and structure (architecture) of a system. Over the years, other terms have also been applied using the DSM initials (e.g., dependency structure matrix, dependency system model, deliverable source map, and other combinations of such words). Most of these alternative terms sprang from a desire to emphasize a particular aspect of a DSM model, such as its ability to model or map dependencies between elements in a system.

In this book, we adopt the name DSM as a unifying term for a wide variety of square matrix models, even though some of these predate the term DSM. Our main criterion is that DSM is a square matrix, with the rows and columns identically labeled and ordered, and where the off-diagonal elements indicate relationships between the on-diagonal elements. Later, in chapter 8, we describe a rectangular matrix called a *domain mapping matrix* (DMM) used to link DSM matrices across domains. Many matrix-based methods have their origins in a branch of mathematics called graph theory, which has tended to focus on analytical techniques. Insights from graph theory continue to provide an important source of extensions to DSM analysis methods.

System Architecture

A *system* is "a combination of interacting elements organized to achieve one or more stated purposes" (INCOSE 2007, p. C5). IEEE (2000) defined *system architecture* as "the fundamental organization of a system embodied in its components, their relationships to each other, and to the environment, and the principles guiding its design and evolution" (p. 3). We adopt this IEEE definition with some minor modifications, mainly for the sake of clarity and emphasis. First, we replace the word "organization" with the generic term "structure," retaining the former for particular application to what are widely referred to as organizations (i.e., assemblies of people). Second, we generalize the product-oriented definition by using the generic term "elements" for any kinds of "components," reserving the latter term for the elements of a product. Third, we make the connection between architecture and function explicit. We therefore use the following definition:

System Architecture: The structure of a system—embodied in its elements, their relationships to each other (and to the system's environment), and the principles guiding its design and evolution—that gives rise to its functions and behaviors.

Thus, a system's architecture describes its elements and their relationships as a structure that can be designed and may evolve over time. We could refer to DSM as "the system architecture design matrix," but "design structure matrix" is simpler, has become widely accepted, and will suffice when understood in the proper context.

All types of systems have architectures. *Product architecture* refers to the components and interactions within a physical artifact, such as hardware (and sometimes software), including automobiles, aircraft, buildings, ships, computers, equipment, machinery, and so on. *Organization architecture* refers to the people or teams and their interactions within an organization. *Process architecture* refers to the actions and interactions that accomplish work, such as the design or production of a product, the delivery of a service, or the execution of software code. While products, organizations, and processes are each a type of system, at times the term "system" is used to refer to any one of these (e.g., complex products or portions thereof are often called systems), and sometimes it is used to refer to all of them collectively. In this book, we strive to use the terms "product," "process," or "organization" to refer to each of these particular types of systems while reserving the more general term "system" for remarks pertaining to any type of system.

Two other categories of relationships that are particularly important in system modeling are hierarchical (vertical) and lateral (horizontal). Hierarchical relationships stem from the *decomposition* or breakdown of a system into elements. For large and/or complex systems, decomposition may recur through several levels (Simon 1962, 1996). Lateral relationships stem from interactions between elements, such as flows of material or information, at the same level. Hierarchical relationships are often modeled with breakdown structure diagrams—for example, work breakdown structures (WBS), organization breakdown structures (OBS) or org charts, and product breakdown structures (PBS) or product trees or indented bills of materials. While a DSM is mainly used to represent the lateral relationships between elements at a particular level of decomposition, it can also show elements' locations in a hierarchy, as illustrated in figure 1.6. Note that the DSM in

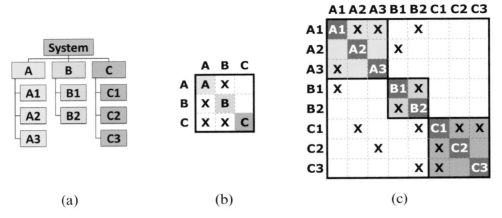

Figure 1.6
Decomposition can be represented with a tree diagram (a) or with a DSM, either at a high level (b) or at a lower level (c).

figure 1.6c shows the lateral relationships among elements at the lowest level of the hierarchy, whereas the DSM in figure 1.6b shows only the presence of these relationships between higher level elements in the hierarchy. Note also that panels a–c of figure 1.6 are not entirely equivalent because the breakdown structure (a) diagram does not include the lateral relationships.

Why are we so interested in system architecture? Simply put, architecture drives behavior. The structure of a system's elements and interactions causes the emergence of system attributes, functions, and behaviors (some anticipated and some not). Architecture also governs a system's performance and value (both short and long term). System designers make choices about elements and relationships. Some designs are better than others. Although many reasons can account for the differences, a key one has to do with skill in developing the system's architecture, which is largely determined by the choices made by designers (or architects) early in the system's development process.

Although the choice of elements to include in a system has always been a focus of system designers, relatively recent advances in complexity science have emphasized the critical role played by the lateral links among elements, particularly when it comes to the emergence of system behaviors. For a system, the value of the whole is greater than the sum of its parts. Or, as Rechtin (1991) explained in the context of designing large, complex systems, "Relationships among elements are what give systems their added value," and, therefore, "The greatest leverage in system architecting is at the interfaces" (p. 29).

As many examples presented in this book demonstrate, it is often possible to make drastic improvements to a system without significantly changing its elements or their interactions. Large benefits can be achieved merely by changing the way the elements are structured—for example, by grouping product components into a different set of modules, by grouping people into a different set of teams, or by altering activity sequences in a process. These kinds of improvements may allow us to better implement the product architecture, more effectively manage the organization, or more efficiently execute the process. Such benefits are often the result of a partitioning analysis, such as sequencing or clustering, applied to the system architecture represented in a DSM format, as illustrated in figure 1.5.

Advantages of the DSM for System Architecture Modeling

Although DSM is the focus of this book, it is important to remember that DSM is only one important tool in a system designer's or modeler's tool kit. In many cases, it is not a question of finding or choosing a single best method, tool, or representation for architectural modeling; rather, a combination of representations is most powerful (Browning 2009). However, within the suite of potential representations, DSM offers some salient advantages:

- **Conciseness**   The structured arrangement of elements and interactions provides a compact representation format. Compared with many other network modeling approaches, we find that a DSM can meaningfully represent a fairly large, complex system in a relatively small space.

- **Visualization**   The DSM highlights relationship patterns of particular interest to a system designer. For example, a process architecture DSM can distinguish feedback interactions, which have magnified implications for the system's behavior, and a product architecture DSM may show regions of heavy interaction indicative of benefits of assigning particular components to subsystems or modules. Moreover, the DSM provides a system-level view that can support more globally optimal decision making and help orient those focused on particular elements.

- **Intuitive Understanding**   Once introduced to DSM, people find that they are able to understand the basic structure of a complex system quickly once the DSM model is properly displayed. Hierarchy and complexity become apparent in even a cursory review of the DSM.

- **Analysis**   The matrix-based nature of the DSM opens the door to applying a number of powerful analyses in graph theory and matrix mathematics as well as specialized DSM analysis methods. DSM analysis can also illuminate indirect links, change propagation, process iterations, convergence, modularity, and other important patterns and effects.

- **Flexibility**   DSM is a highly flexible system modeling tool. Since its initial development more than three decades ago, many researchers and practitioners have modified and extended the basic DSM with helpful graphics, colors, and additional data. New possibilities continue to develop every year.

In our experience of building and evaluating DSM-based models, we have found that merely building the model provides several important benefits. Modeling may prompt the acquisition of previously latent system information and stimulate dialog among various experts, which serves to increase the alignment of their individual mental models. The DSM provides a common perspective on a system, increases designers' understanding of the cause-and-effect relationships occurring within the system, helps to organize this knowledge, and channels creativity and innovation toward beneficial improvements—in other words, DSM helps people better manage system complexity.

DSM Approach to Architectural Modeling and Analysis

Each of the DSM applications presented in this book essentially follows a five-step approach to architectural modeling and analysis (figure 1.7). These steps are:

1. **Decompose**   Break the system down into its constituent elements perhaps through several hierarchical levels.

2. **Identify**   Document the relationships among the system's elements.

3. **Analyze**   Rearrange the elements and relationships to understand structural patterns and their implications for system behavior.

4. **Display**   Create a useful representation of the DSM model, highlighting features of particular importance or of special interest.

5. **Improve**   Most DSM applications result in not only better understanding of the system but also improvement of the system through actions taken as a result of the DSM analysis and interpretation of its display.

These five steps should be prefaced by appropriate preparation and planning in light of the purposes, goals, and constraints for the effort. We do not advance these steps as sufficient to completely understand a complex system; rather, we submit that they can be extremely powerful and beneficial when designing and managing engineering and socio-technical systems. Ideally, the last step would include a feedback to the first, closing not only a continuous improvement loop but also a systematic cycle of learning that increases knowledge of the system and improves the model's accuracy and richness.

Users of DSM may be process owners, project managers, or consulting staff who are interested in exploiting the insights DSM can provide to improve the system. The DSM

Figure 1.7
The DSM approach to system modeling, analysis, and improvement.

approach outlined here generally takes practitioners several weeks to execute. The actual time and effort required for any DSM application would depend on several factors, including familiarity with the system being modeled, access to product and process experts, available documentation, level of modeling detail desired, and experience of the DSM modelers. Although some DSM models can be extracted automatically from project management models or software code, most have entailed the direct involvement of experts.

Types of DSM Models

Almost all DSM models to date may be classified into four types within three main categories as shown in figure 1.8. The first category consists of *static architecture* models, representing systems whose elements exist simultaneously. Types of applications in this category include systems such as products (whose components physically interact with one another) and organizations (whose members communicate with one another). The second category consists of *temporal flow models*, representing systems whose elements may be actuated over time. All of the applications in the temporal category are types of

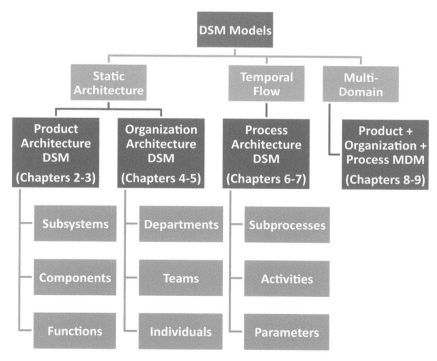

Figure 1.8
Four types of DSM models. Each type is the focus of two chapters of this book, one chapter introducing the modeling method and one chapter providing several industrial application examples.

processes, represented as activity-based process models, low-level parameter-based models, and even software processes (although software is a product, it executes procedurally). The third category consists of *multidomain matrix* (MDM) models, which represent more than one type of DSM (e.g., product, process, and/or organization) in a single matrix. We present methods for and examples of each of these types of DSM models in this book.

Many readers will note that hardware products and organizations certainly have temporal properties, and there might be insights in applying a temporal DSM to these systems (Browning 2001), but here we focus on classifying existing applications. So far, studies of product and organizational evolution have employed a collection of static DSMs as "snapshots" in time (e.g., Sosa et al. 2007 and examples 5.2 and 9.12).

These four types of DSM models have led to DSM applications in many industries addressing a wide range of problems and situations. Because the primary, single-domain applications of DSM have been products, organizations, and processes, we organize the sections of this book on that basis. Hence, we have grouped all of the temporal flow DSM models into the process DSM type. A fourth section of the book presents the multidomain models and their application.

Although we focus this book on DSM research and applications to date in complex, engineered systems, we hasten to point out that many other applications have great potential, including portfolios of projects or investments, supply chains or networks, information technology platforms, enterprise goals and objectives, product requirements, systems of risk, product-service systems, public policy, and social systems, to name but a few.

A Brief History of DSM

Before the term DSM was coined by Professor Don Steward of California State University, Sacramento, in the 1970s, a branch of graph theory had long used square precedence matrices to depict relationships among nodes in a digraph. However, Steward has received primary credit for creating the DSM method by first applying the square-matrix format to represent a network of design variable (or design task) interactions. The technique was derived from methods used to sequence large systems of equations in order to solve them with minimal iteration. In applying this approach to design variables and processes, he explained that the benefits of the DSM method would include "to develop an effective engineering plan, showing where estimates are to be used, how design iterations and reviews are handled, and how information flows during the design work" (Steward 1981b). Other contemporary diagramming methods used in the 1970s included process flow charts, N^2 charts, and node-link diagrams, all of which would eventually be transformed into various DSM formats.

At MIT, we picked up DSM in 1989, recognizing its potential in Steward's book (Steward 1981a) and *IEEE Transactions* paper (Steward 1981b). However, we wondered why there did not appear to be any industrial application of the method. With a series of master's and doctoral students, we undertook a number of industry projects in the automotive, electronics, and aerospace industries in the 1990s (Eppinger et al. 1990, 1994). Through this experience, we found both the challenge and the promise of DSM for complex systems industries. We extended DSM from Steward's initial process flow models and sequencing analysis to include static architectural models, clustering analysis, and a range of applications to product and organization domains.

Early industry application (and further development) of DSM began at NASA, Boeing, General Motors, and Intel in the early 1990s. (Several of these works are cited in later chapters.) Today, there are applications of DSM spanning many more firms in a range of industries, as partly evidenced by the examples provided in this book.

The DSM research community was established in the late 1990s with a series of workshops held at MIT. This worldwide DSM community now includes researchers at universities throughout Europe, Asia, Australia, South America, and North America. We also include in the DSM community a network of software developers, consultants, and lead users in industries where DSM is being actively applied. Indeed, many of the examples shown in this book come from members of this growing DSM community.

By now there are hundreds of research papers that chronicle the development of DSM methods and document a wide variety of applications. An extensive listing of these papers can be found on the DSM community website (www.dsmweb.org).

Structure of This Book

We have organized the remainder of this book in four parts corresponding to the four primary types of DSM application described earlier. Each part contains two chapters. First, a chapter describes how the DSM modeling approach is applied to the particular domain, how such models are analyzed, and the kinds of industrial problems for which we have found DSM to be useful. Second, a chapter presents a variety of application examples in a range of industries. Many researchers and practitioners have contributed examples to this book, and we provide references to their original publications whenever possible, as their presentations in this book are necessarily abbreviated.

We start with static DSM models in the product domain in chapter 2, with our attention focused on product architectures, followed by example applications in chapter 3. We then apply DSM in the organization domain in chapter 4, with a focus on the structure of and communications within organizations, followed by example applications in chapter 5. Chapter 6 presents DSM in the process domain, adding a sequential orientation and temporal flow to the DSM, followed by example applications in chapter 7. Finally, chapter 8 presents several types of multidomain models, combining types of DSM models seen in the earlier chapters to represent multiple domains simultaneously, followed by example applications in chapter 9. We conclude in chapter 10 with a look at the future of DSM methods and applications.

References

The introductory volume of this *MIT Press Series on Engineering Systems* offers a comprehensive overview of the challenges and opportunities of complex engineered systems today.

de Weck, Olivier L., Daniel Roos, and Christopher L. Magee. 2011. *Engineering Systems: Meeting Human Needs in a Complex Technical World.* Cambridge, MA: MIT Press.

Fundamental work on the architecture of complex systems began in the 1960s.

Alexander, Christopher. 1964. *Notes on the Synthesis of Form.* Cambridge, MA: Harvard University Press.

Rechtin, Eberhardt. 1991. *Systems Architecting: Creating & Building Complex Systems.* Englewood Cliffs, NJ: PTR Prentice-Hall.

Simon, Herbert A. 1962. The Architecture of Complexity. *Proceedings of the American Philosophical Society* 106 (6):467–482.

Simon, Herbert A. 1996. *The Sciences of the Artificial.* 3rd ed. Cambridge, MA: MIT Press.

IEEE and INCOSE provide helpful background on systems architecting and systems engineering definitions and methods.

IEEE. 2000. *IEEE Recommended Practice for Architectural Description of Software-Intensive Systems.* Institute of Electrical and Electronics Engineers Standards Association, IEEE Std 1471–2000.

INCOSE. 2007. *Systems Engineering Handbook: A Guide for System Life Cycle Processes and Activities,* Version 3.1, International Council on Systems Engineering (INCOSE).

Steward's original 1981 book and *IEEE Transactions* paper first explained basic DSM methods and applications to sequencing design parameters.

Steward, Donald V. 1981a. *Systems Analysis and Management: Structure, Strategy, and Design.* Princeton, NJ: Petrocelli Books (original edition out of print, but reprinted by TAB Books, 1997).

Steward, Donald V. 1981b. The Design Structure System: A Method for Managing the Design of Complex Systems. *IEEE Transactions on Engineering Management* 28 (3):71–74.

DSM research at MIT expanded Steward's original focus from design parameters to processes, organizations, and product architectures. The following articles summarize some of these efforts, while many of them appear as separate examples and references in later chapters.

Eppinger, Steven D., Daniel E. Whitney, Robert P. Smith, and David A. Gebala. 1990, September. Organizing the Tasks in Complex Design Projects. *ASME Conference on Design Theory and Methodology,* Chicago, IL, pp. 39–46.

Eppinger, Steven D., Daniel E. Whitney, Robert P. Smith, and David A. Gebala. 1994. A Model-Based Method for Organizing Tasks in Product Development. *Research in Engineering Design* 6 (1):1–13.

In particular, Browning reviewed many applications of DSM (and closely related modeling methods), establishing the distinction between static and temporal DSMs (and their associated analysis techniques, clustering, and sequencing, respectively) and the categories of product, organization, process, and parameter DSMs. He also discussed the potential for crossover applications (e.g., sequencing a product DSM or clustering a process DSM) and multidomain analyses.

Browning, Tyson R. 1998. *Modeling and Analyzing Cost, Schedule, and Performance in Complex System Product Development.* PhD thesis (TMP), Massachusetts Institute of Technology, Cambridge, MA.

Browning, Tyson R. 2001. Applying the Design Structure Matrix to System Decomposition and Integration Problems: A Review and New Directions. *IEEE Transactions on Engineering Management* 48 (3):292–306.

The DSM can be used in conjunction with other types of charts, diagrams, and representations (views) to provide a rich, multifaceted model of a system.

Browning, Tyson R. 2009. The Many Views of a Process: Towards a Process Architecture Framework for Product Development Processes. *Systems Engineering* 12 (1):69–90.

Sosa et al. used a longitudinal set of static DSMs to analyze the dynamics of evolving products.

Sosa, Manuel E., Tyson R. Browning, and Jürgen Mihm. 2007, September 4–7. *Studying the Dynamics of the Architecture of Software Products.* Proceedings of the ASME 2007 International Design Engineering Technical Conferences & Computers and Information in Engineering Conference (IDETC/CIE 2007), Las Vegas, NV.

2   Product Architecture DSM Models

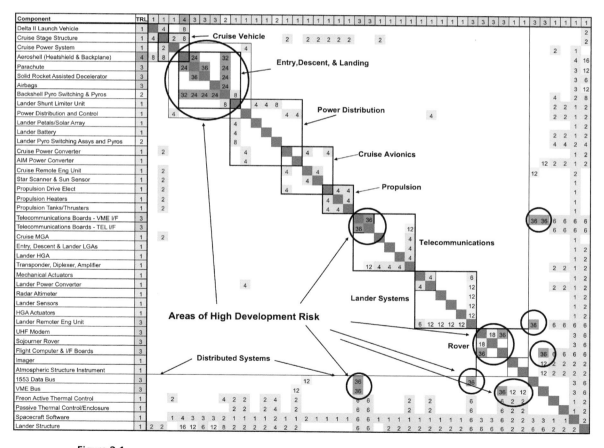

Figure 2.1
A product architecture DSM model augmented to represent technology risks in NASA's Mars Pathfinder program (see example 3.4).

In this chapter, we turn our attention to the architecture of complex products. We show how DSM is applied to represent and analyze these architectures and the types of insights gained through such applications. We begin with a brief synopsis of terminology used in the particular context of product architecture DSM modeling.

Terminology

Product or System   A complex product or engineered system. Such systems come in many forms and include automobiles, aircraft, electronics, software, mechatronics, machinery, capital equipment, built environments, etc. In this chapter, the general term "system" refers to either the product itself or the product and its surrounding environment or supporting infrastructure.

Product Architecture   The arrangement of components interacting to perform specified functions. The architecture of a product is embodied in its components, their relationships to each other and to the product's environment, and the principles guiding its design and evolution. The terms **product architecture** and **system architecture** are used interchangeably in certain contexts.

Components   The elements comprising a product. Depending on one's point of view, a component may be a complex product or system.

Interactions   The relationships between components or elements in a system. Interactions may be of various types depending on the nature of the system. Many interactions occur through interfaces between components.

Product Architecture DSM   A mapping of the network of interactions between a product's components, also known as **system architecture DSM**, **product DSM**, and **component-based DSM**.

Cluster   A set of components grouped because of certain relationships, suggested through analysis of the product architecture DSM, and defined to comprise a module or subsystem.

Background

In defining the architecture of complex products and systems, it is common to decompose the product or system into smaller elements such as subsystems, modules, and components. These elements must be integrated to work together in order to achieve the performance of the system as a whole. The field of systems engineering is largely concerned with delivering system-level performance by planning and controlling the network of interactions between components and subsystems. The traditional *systems engineering V diagram*, shown in figure 2.2, illustrates the process of developing complex systems through design and decomposition on the "down side of the V" and through component-to system-level integration and testing on the "up side of the V."

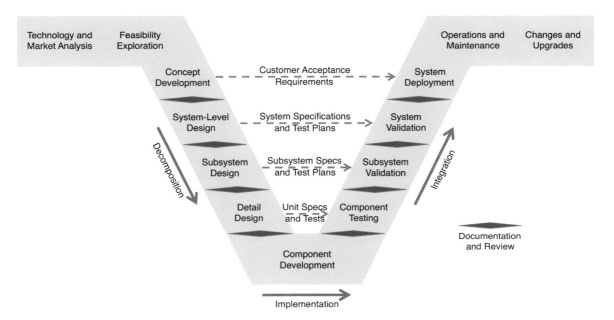

Figure 2.2
Systems Engineering V (adapted from U.S. Department of Transportation).

According to our definition, a product's architecture has to do with the way its components work together to perform its functions. Developing the architecture involves three mappings: (1) hierarchical decomposition of the product into modules and components—often represented by a product breakdown structure diagram, (2) assignment of functions to the modules and components—sometimes mapped using a rectangular matrix diagram, and (3) interactions between modules and components—the focus of our DSM applications in this domain.

DSM research in the product domain has been motivated by two primary objectives: design of superior architecture (down side of the V) and improved implementation of the architecture through more effective system integration (up side of the V). Important advantages derived from improved architecture and integration may include: enabling decomposition and segmentation of a product's associated development process and organization structure (which in turn facilitates outsourcing and project management), guiding standardization of internal interfaces, facilitating integration and testing at the component and system levels, addressing potential quality problems during product development, enabling product platforms and families with appropriate percentages of common components, and reducing the costs of product adaptation and redesign. Of course, these are in addition to the general advantages for complexity management, visualization, understanding, and innovation mentioned in chapter 1.

DSM has been used by a number of researchers and practitioners for product architecture analysis. Depending on the context or author, these DSMs have been given many different names, including *product architecture DSM*, *system architecture DSM*, *product DSM*, and *component-based DSM*. In all of these cases, this type of DSM model represents the components comprising a product and the relationships between them.

While node-link graph models go back much further, to our knowledge, the first instance of a square matrix being used to represent a system's components and their relationships is what systems engineers call an N-square (N^2) chart or diagram. The first written source on N^2 diagrams that we are aware of is Lano's 1977 TRW report, later published as a book (Lano 1979). However, it is our understanding that N^2 diagrams have been in use internally by various U.S. aerospace companies since perhaps the 1950s or 1960s (along with architecture block diagrams and entity-relationship diagrams, which may show similar content in more of a flowchart format). Figure 2.3 shows an example of an N^2 diagram, which is similar to the DSM (in this case, using the inputs in columns [IC] convention). This use of square matrices to model system interfaces continues in the systems engineering community, notably through inclusion in architecture frameworks such as the U.S. Department of Defense Architecture Framework (DoDAF) (DoD 2009). In the 1980s, the House of Quality (Hauser and Clausing 1988), with its "roof" comprising a triangular half of a square matrix, and Quality Function Deployment (QFD) (e.g., Akao 1990) also demonstrated some of the benefits of mapping the relationships between product elements.

In 1994, researchers at MIT published a DSM model (figure 2.4 and described more fully in example 3.1) representing a product's architecture as a network of components and their interactions (Pimmler and Eppinger 1994). This research exposed benefits of distinguishing different types of interactions among components (such as spatial proximity, material flow, information flow, and energy transfer) and of analyzing the model to prescribe alternative architectures with improved modularity. Since then, the use of square matrices to model product architectures has continued, and many (but not all) of these applications have used the term DSM.

Using product architecture DSM models, many researchers and industrial practitioners have been able to better understand networks of interactions in complex systems, yielding two primary types of benefits:

- **Architecture benefits** Planning subsystems or modules, understanding connections across subsystems or modules, identifying the impact of new technology, assessing the match between technical and organizational architectures, designing for modularity, designing for adaptability

- **Integration benefits** Planning necessary integration and test activities at component, module, and subsystem levels; identifying problematic interactions that may present integration challenges

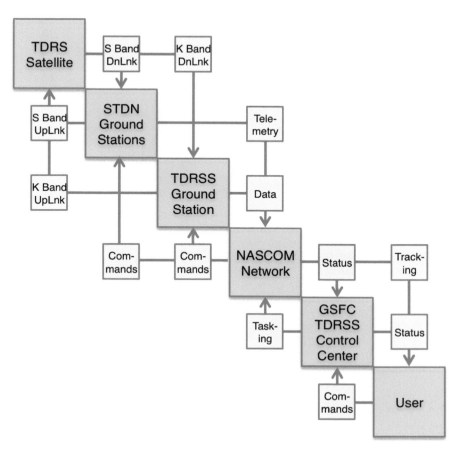

Figure 2.3
N^2 diagram from early systems engineering (adapted from Lano 1979, p. 96).

Building a Product Architecture DSM

The basic procedure for building a product architecture DSM is as follows:

1. Decompose the overall product or system into its subsystems and/or components. Lay out the square DSM with components labeling the rows and columns, grouped into subsystems or modules if appropriate.

2. Identify the known interactions between the components and represent these using marks or values in the DSM cells.

The climate control DSM model in figure 2.4 illustrates this basic procedure. The system is decomposed into 16 components, represented by the 16 × 16 DSM. Interactions are

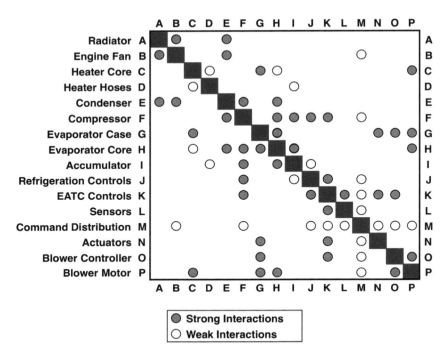

Figure 2.4
Product architecture DSM model of an automobile climate control system.

shown to be either strong or weak. (The original work used four dimensions to document four types of interactions, but we summarize them here using marks denoting only two levels of strength.)

Here are several caveats to consider when creating product architecture DSM models:

- **Boundaries**   The limits of the designated system may or may not be well understood. Choose the system boundaries so as to include all of the relevant components and interactions you would like to represent in the DSM model. Early drafts of the model may prompt one or more revisions of the system boundary as it becomes apparent that including or excluding certain components will make the model more useful. (An interesting example of this caveat is provided by the climate control system DSM, in which the vehicle engine is outside the DSM system boundary. See the discussion of this concern in example 3.1.)

- **Interaction types**   Consider the various types of interfaces, relationships, and interactions that may exist among components. Some interactions may be well defined, such as physical adjacency of mating parts or flow of materials among subsystems. Other interactions may be poorly understood, hidden, or only occurring under certain conditions, such as heat transfers, vibrations, electrical interferences, or environmental effects. Different marks, values, or colors in the DSM cells may be used to indicate the various types of interactions.

- **Interaction strengths**   Consider the level, strength, or degree of interaction among the components. Instead of a binary DSM, a numerical DSM may be used to show varied levels of interaction in the off-diagonal cells; for example, a simple scale such as weak or strong or a numerical scale with additional levels of distinction could be used. Positive and negative values could distinguish desirable from undesirable interactions.

- **Symmetry**   Most interactions in product architecture DSMs are symmetric. That is, if component A interacts with component B, then B also interacts with A. Asymmetric interactions can also be present depending on the types of interactions in the model. For instance, component C may create noise, which affects component D but not vice versa.

- **Granularity**   There is a tradeoff between greater richness of the model by decomposing into smaller components versus modeling simplicity and ease of interpretation by limiting the model's granularity. We usually find there is a sweet spot representing the right compromise here, so we recommend starting small, with a manageable DSM size of 20–50 components, and only adding components and interactions where additional richness is needed. Here it is especially important to keep in mind the purpose of the model, such as supporting a specific decision, and the available resources for building the model, all of which will help determine the appropriate (or feasible) amount of detail. Subsequent analysis of the DSM may reveal portions of the model where additional granularity is needed as well as portions that could be rolled up into lesser detail without much loss of information or insight.

- **Identifying interactions**   Interaction data for the DSM may come from product documentation, interface specifications, and the like. However, for most product DSM models, the data collection requires at least some amount of direct discussion with subject matter experts in order to draw out the tacit and system-level knowledge that may not be captured in the documentation. Experts should also be consulted to verify and validate the model.

Successful DSM models tend to meet the following criteria:

- The models have a clear purpose (not modeling for the sake of modeling).
- The models use the appropriate amount of detail for the intended purpose.
- The modelers have access to sufficient knowledge or expertise regarding the system.
- The DSM is maintained as a "living model," continuously improving it by incorporating new knowledge as it becomes available.
- Having a DSM model often prompts the emergence of otherwise latent knowledge.

It is worth noting that, although most of the discussion in this chapter pertains to DSM models of hardware products, software product architecture can also be represented with a DSM, where the components are generally modeled at the level of subroutines, functions, or class files and the interactions are data flows and/or function calls. However, because software actually executes as a process, it is also useful to apply the techniques of process DSMs (chapter 6). Hence, we provide examples of software product DSMs in both chapters 3 and 7 (see examples 3.5 and 7.15). Each type of DSM model and analysis provides a different set of insights.

Analyzing the Product Architecture DSM

Quite a bit of useful insight can be gained merely by building a product architecture DSM model. Many further insights can be derived through careful analysis of the model. The most common method of analysis applied to product architecture DSM models is called *clustering*. This is a form of partitioning analysis that reorders the rows and columns of the DSM to group the components according to some objective, which usually pertains to the number and strength of the interactions. Clusters may be formed to group components that may achieve efficiencies through common membership in the cluster. For example, several components produced by a common supplier, sharing multiple interfaces, or having complex interactions may be candidates for a cluster.

One of the prominent heuristics in systems architecting is to choose modules such that they are as independent as possible (i.e., modules with relatively few external interactions and relatively more internal ones) (Rechtin 1991). However, it is quite common in complex systems to have both modular and integrative subsystems, as explained in a paper by Sosa, Eppinger, and Rowles (2003).

Figure 2.5 shows the result of clustering the climate control system DSM. This DSM analysis indicates three groups of components with many strong, intragroup interactions and relatively few intergroup interactions. The groups are labeled Front-End Air, Refrigerant, and Interior Air. Such groups have been called clusters, chunks, subsystems, or modules by various authors and in various contexts. The clustering result also identifies five highly integrative components within the climate control system, forming a distributed cluster labeled Controls/Connections. A fuller explanation of the climate control system example and additional clustering results is given in the next chapter (example 3.1), along with a range of additional examples illustrating clustering in a variety of applications and showing several modifications to the basic approach.

Clustering is essentially a type of assignment problem seeking the optimum allocation of the N components to M clusters. Clustering algorithms have many applications besides the DSM (e.g., portfolio and market segmentation), and a variety of algorithms are available (e.g., Hartigan 1975). However, a DSM clustering analysis presents several potential challenges.

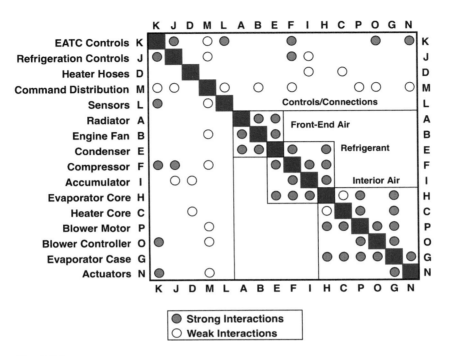

Figure 2.5
Clustered climate control system DSM model.

Clustering objective functions for DSM analysis trade off two conflicting goals: (1) minimize the (number and/or strength of) interactions outside clusters, and (2) minimize the size of the clusters. The determination of the best objective functions for various DSM clustering applications is an important area of ongoing research. (The Reference section at the end of this chapter points to some DSM clustering approaches.) Nevertheless, comparing the clusters obtained by several different objective functions can often lead an analyst to useful insights about the product architecture.

Figure 2.6 provides an illustration of clustering analysis based on a simple objective function. For this example, we used a portion of the climate control system DSM (based on only the materials interactions, as explained in example 3.1). In this illustration, we show four possible clustering solutions, with two or three clusters, with or without overlapping. The objective function to be minimized considers both the size of the clusters (C_i) and the number of interactions outside the clusters (I_o), according to the following equation, where $\alpha = 10$ and $\beta = 100$:

$$Obj = \alpha \sum_{i=1}^{M} C_i^2 + \beta I_o$$

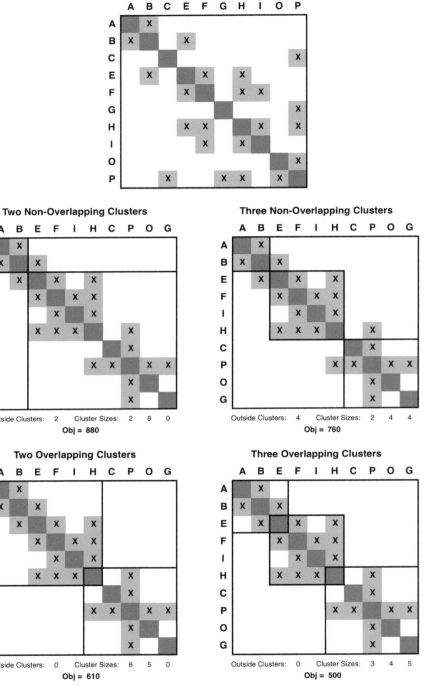

Figure 2.6
Clustering analysis based on a simple objective function to minimize both the size of clusters and the number of interactions outside the clusters.

Clustering analysis also requires attention to the following considerations:

- **Number of clusters**   What should be the bounds on M? Without any bounds, an objective function might find it optimal just to call the whole DSM a single cluster ($M = 1$) or to call each component a separate cluster ($M = N$), although neither of these extreme solutions is typically desirable. The analyst can gain insight by comparing the different solutions found while specifying varied numbers (or ranges) of clusters as a constraint.

- **Cluster size**   A related consideration is if and how to bound the size of each cluster. Usually, a lower bound of a cluster consisting of a single component should be allowed. However, it may be necessary to constrain the maximum number of components that can be assigned to a cluster. Allowing the size of clusters to increase essentially limits the maximum number of clusters.

- **Overlapping clusters**   The clusters overlap in figure 2.5 and also in two of the clustering solutions shown in figure 2.6. However, most non-DSM clustering algorithms do not support a component's membership in two or more clusters. Nevertheless, the identification and highlighting of such linking components provides an important architectural insight. Therefore, DSM clustering analysis generally allows for the possibility of cross-membership in the clusters if such solutions are of interest in the particular case (e.g., Yu et al. 2007).

- **Interaction types**   We also mentioned earlier that the model in figure 2.5 actually accounts for four different types of interactions among the components. (These are spatial proximity, material flow, information flow, and energy transfer, as presented in further detail in example 3.1.) Clustering the DSM based on any one of these types of interactions alone is likely to suggest a different set of clusters than clustering on the combination of interactions. This begs the question of whether certain types of interactions should be weighted more heavily than others by the objective function. For example, spatial proximity interactions might be more difficult to achieve via a standard interface than information flow interactions, which might be more amenable to a standard protocol. Again, the analyst can often gain useful insights by comparing the different optimal solutions found when considering the different types of interactions collectively and separately.

- **Integrating elements**   Figure 2.5 shows a number of components in the upper left of the DSM that are not explicitly assigned to a cluster because they each have significant interactions with many components. These integrative or bus components often serve collective functions (such as monitoring and control) or as interaction conduits (e.g., the hoses in the climate control system example or a literal bus or backplane in an electronic system). Hence, some clustering algorithms allow the analyst to set a bus threshold, an amount of interaction above which a component is assigned to an integration cluster (for which it would be sorted toward the upper left or lower right corner of the DSM). As with other considerations discussed previously, the analyst can gain insight through sensitivity analysis of the bus threshold.

- **Manual clustering**   Although automated clustering methods are available in software programs, many product architecture DSMs can be analyzed directly by moving the rows and columns manually (or with programmed macros) in a spreadsheet application or by using the manual sorting functions provided in most DSM software. Manual adjustment is also useful for sensitivity analysis around the solutions proposed by clustering algorithms.

- **Multiple clustering solutions**   Because modularization involves balancing so many factors, we find it useful to suggest several solutions and consider the interpretation of each cluster before accepting any results.

Applying the Product Architecture DSM

Product architecture DSMs have been applied to a range of industrial problems and have produced many useful insights. Several examples are given in the next chapter. Typical applications include:

- Enhancement of product modularity, which determines subsystem boundaries, relates to component sharing across product lines, and affects the difficulty of outsourcing and system integration (see examples 3.2, 3.4, 3.5, 3.9).

- Carefully scrutinizing the clusters suggested by DSM analysis and comparing these to established subsystems, subassemblies, or modules.

- Identification and application of "design rules," which systems architects and engineers use to guide and enforce standards across the product architecture (Baldwin and Clark 2000).

- Using insights from the product architecture to inform the design of the product development process and/or organization. Planning and managing the system integration process (the up side of the systems engineering V) based on the network of interactions (see examples 3.1, 3.2, 3.4, 3.6, 3.7, 5.1, 5.3, 7.8, 9.2, 9.4).

- Understanding product architecture dynamics, evolution, and adaptability across multiple generations. Sosa et al. (2007) described some metrics for the number of new, deleted, and changed components and interactions from one product version or generation to the next. Engel and Browning (2008) explored modularization in support of design for adaptability and described an approach where the clustering objective function accounts for each component's option value and change cost, which tends to separate dynamic components from stable ones to facilitate changeability, upgradeability, etc. in the design of product platforms and families (see examples 3.5, 3.6, 3.9).

- Interface management. The DSM can be used to identify and monitor key interfaces. It can be augmented with attributes such as the name and owner of each interface. Each such interface could even spawn a formal interface control document (see examples 3.1, 3.4, 3.6).

- Product portfolio management. Product architecture DSMs can be overlaid to determine common and variant components in a product portfolio or family (see examples 3.9, 9.1).

- In addition to modeling product architectures—and, in chapters 4–5, organization architectures—static DSMs also have the potential to be used more broadly for applications in many other nonengineering domains such as public policy analysis and portfolio segmentation/diversification in financial products.

Conclusion

The product architecture DSM provides a highly effective representation for product components and their relationships. It documents both the product decomposition and the network of interactions. It can be analyzed via clustering analysis, which (although it remains somewhat of an art) generates alternative groupings of components into modules, improves architectural understanding, and facilitates architectural innovation.

The value of the product architecture DSM increases as products become larger and more complex systems. This is because system complexity makes it impossible for any single individual to have a complete, detailed, and accurate mental model of the entire system. The DSM helps individuals to communicate, compare, and integrate their partial models of the system. Indeed, two of the main benefits of a DSM model are its abilities to (1) concisely represent a relatively large number of components and their relationships, and (2) highlight important groups of components and patterns of interactions, such as those influencing modularity.

References

This list of references provides additional background on the product architecture DSM.

Lano produced an internal report at TRW in November 1977 titled "The N^2 Chart," published as a book in 1979, which described the mechanics of the N^2 diagram (see figure 2.3) for a variety of applications similar to the DSM but in a more graphical format than a matrix. Whereas systems engineers continue to use N^2 diagrams, DSM encompasses most of its capabilities while adding many analytical benefits.

Lano, R. J. 1979. *A Technique for Software and Systems Design.* New York: North-Holland.

Pimmler and Eppinger developed the first application of DSM to product and system architecture, representing component-to-component interactions (see example 3.1). (Prior DSM work had been limited to process- and parameter-based models.)

Pimmler, Thomas U., and Steven D. Eppinger. 1994, September. *Integration Analysis of Product Decompositions*. Proceedings of the ASME International Design Engineering Technical Conferences (Design Theory & Methodology Conference), Minneapolis, MN.

The following sources provide insights on approaches and algorithms for clustering.

McCormick, William T., Paul J. Schweitzer, and Thomas W. White. 1972. Problem Decomposition and Data Reorganization by a Clustering Technique. *Operations Research* 20 (5):993–1009.

Hartigan, John A. 1975. *Clustering Algorithms*. New York: John Wiley & Sons.

Fernandez, Carlos Iñaki Gutierrez. 1998. *Integration Analysis of Product Architecture to Support Effective Team Co-Location*. Master's thesis (ME), Massachusetts Institute of Technology, Cambridge, MA.

Thebeau, Ronnie E. 2001. *Knowledge Management of System Interfaces and Interactions for Product Development Processes*. Master's thesis (Eng. & Mgmt.), Massachusetts Institute of Technology, Cambridge, MA.

Yu, Tian-Li, Ali A. Yassine, and David E. Goldberg. 2007. An Information Theoretic Method for Developing Modular Architectures using Genetic Algorithms. *Research in Engineering Design* 18 (2):91–109.

Hölttä-Otto, Katja, V. Tang, and Kevin Otto. 2008. Analyzing Module Commonality for Platform Design using Dendrograms. *Research in Engineering Design* 19 (2):127–141.

Zakarian, Armen. 2008. A New Nonbinary Matrix Clustering Algorithm for Development of System Architectures. *IEEE Transactions on Systems, Man and Cybernetics. Part C, Applications and Reviews* 38 (1):135–141.

Baldwin and Clark used DSM models to illustrate the nature of modularity in product architecture and discussed the benefits of modular architectures.

Baldwin, Carliss Y., and Kim B. Clark. 2000. *Design Rules*. Cambridge, MA: MIT Press.

Engel and Browning suggested a clustering objective function to enable architecture options (i.e., modularizing the components with the greatest change cost and option value).

Engel, Avner, and Tyson R. Browning. 2008. Designing Systems for Adaptability by Means of Architecture Options. *Systems Engineering* 11 (2):125–146.

The "roof" of the House of Quality (formally known as Quality Function Deployment) is a triangular mapping of relationships between product attributes. This is essentially half of a DSM model in the domain of customer needs and/or product specifications.

Akao, Yoji, ed. 1990. *Quality Function Deployment*. Cambridge, MA: Productivity Press.

Hauser, John R., and Don Clausing. 1988. The House of Quality. *Harvard Business Review* 66 (3):63–73.

By comparing the density of interactions within versus across subsystems, Sosa et al. explained that the product architecture-based notions of modularity and integrality also apply more generally to the architectures of complex systems (see example 3.2).

Sosa, Manuel E., Steven D. Eppinger, and Craig M. Rowles. 2003, June. Identifying Modular and Integrative Systems and Their Impact on Design Team Interactions. *Journal of Mechanical Design* 125 (2):240–252.

Sharman and Yassine used the DSM to identify several characteristic patterns in product architectures, including modules, chunks, various kinds of buses, pinning, and holding away.

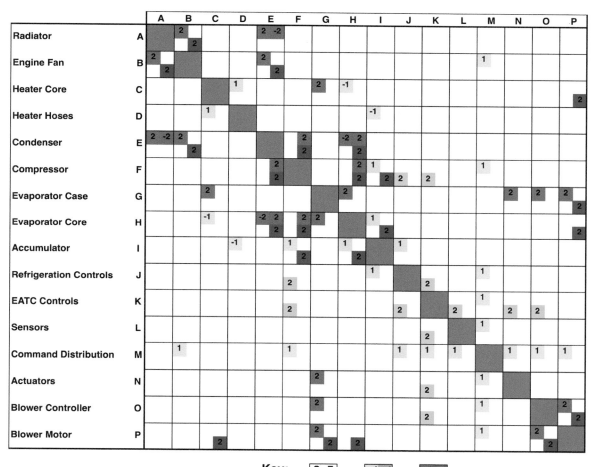

Figure 3.1.1
Composite DSM including interactions among components of four types: spatial, energy, information, and materials.

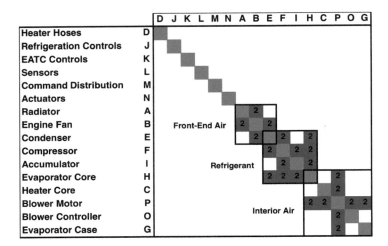

Figure 3.1.2
Clustered materials DSM.

product's architecture. We also showed a way to depict multiple interaction types in each cell and the application of multidimensional clustering analysis.

Three important clusters (interior air, refrigerant, and front-end air) were identified by considering only the materials transfer-type of interactions, as shown in the clustered materials DSM (figure 3.1.2). We also performed clustering analysis with the other three dimensions. A group of integrating components (controls/connections) with interactions across the entire system was primarily identified through analysis of the DSM based on the information-type interactions. Finally, to create the composite clustered DSM, we combined the results of each of the four single-dimensional clustering analyses.

Considering the composite clustering results, we make several observations. The three clusters included interactions of the materials, energy, and spatial adjacency types. However, in the highly integrative controls/connections chunk, the interactions were of the spatial and information types. This suggests that for some systems, certain types of interactions may be clustered as product modules, whereas other interactions are more integrative across the entire product or system.

It is also interesting to note that there is no cluster related to the flow of engine coolant through the radiator and the heater core to provide heating to the passenger compartment. The automobile's engine was not a part of this analysis because Ford did not consider it a climate control component. (Engines were produced by Ford's Powertrain Division.) Without the engine to couple the heating elements, the analysis did not identify the heating loop. An important lesson here is to be careful where to draw the boundaries of the system being represented by the DSM analysis.

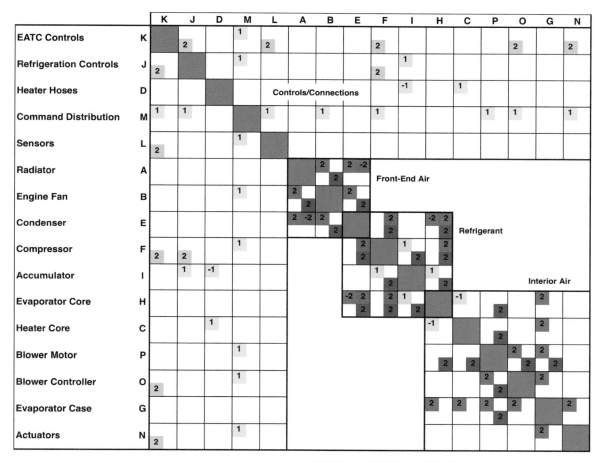

Figure 3.1.3
Clustered composite DSM.

As hoped, this analytical result had organizational implications for CCD. They were considering how best to organize climate control system development, and several groupings of components and teams had already been proposed. This analysis was able to provide an additional, objective perspective on the structure of the system based on data describing the network of interfaces between the components. The subsequent reorganization of CCD partly reflected the clustering results shown here.

The four interaction types used in this analysis seemed appropriate for this application. In general, we would expect that other interaction types might be better suited to representing other types of systems. We also expect that each dimension of interaction analysis could yield some useful insights.

Reference

Pimmler, Thomas U., and Steven D. Eppinger. 1994, September. *Integration Analysis of Product Decompositions*. Proceedings of the ASME International Design Engineering Technical Conferences (Design Theory & Methodology Conference), Minneapolis, MN.

Example 3.2   Pratt & Whitney Jet Engine

Contributors

Steven Eppinger
Massachusetts Institute of Technology

Manuel Sosa
INSEAD

Craig Rowles
Pratt & Whitney

Problem Statement

Pratt & Whitney, a division of United Technologies Corporation, produces and supports aircraft jet engines, industrial gas turbines, and space propulsion systems. Development of a commercial aviation jet engine is a highly complex process involving hundreds of engineers working simultaneously on the various components and subsystems. This DSM application investigated the system engineering and system integration aspects of the PW4098 jet engine development process through a product architecture DSM. The engine, as pictured in figure 3.2.1, is decomposed into eight subsystems, which are comprised of 54 major components.

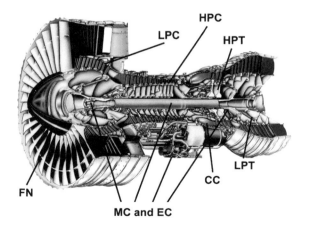

FN: Fan

LPC: Low-Pressure Compressor

HPC: High-Pressure Compressor

CC: Combustion Chamber

HPT: High-Pressure Turbine

LPT: Low-Pressure Turbine

MC: Mechanical Components

EC: Externals and Controls

Figure 3.2.1
PW4098 Jet Engine (courtesy of United Technologies Corp.).

Data Collection

Over a period of four months in 1998, Craig Rowles (both an employee of Pratt & Whitney and a student in MIT's System Design and Management master's program at the time) interviewed system architects responsible for each major component in the PW4098 engine program. To reliably capture as many direct dependencies as possible, he asked about interfaces between components based on known interactions of five types: spatial adjacency, energy flows (e.g., heat), material flows, structural connectivity, and information flows (e.g., data and control signals). Subsequent analysis and interpretation of the DSM model was done jointly with Manuel Sosa, a doctoral student at MIT at the time.

Model

The binary DSM model displayed in figure 3.2.2 shows the decomposition of the PW4098 engine into its eight subsystems and 54 components. Interfaces are indicated in the DSM using red shaded cells between pairs of components.

Results

The DSM model identified six of the subsystems as somewhat modular, in that each subsystem primarily had interfaces among components within the subsystem. These modular subsystems (listed starting from the front of the engine and from the top of the matrix) are the fan, low-pressure compressor, high-pressure compressor, combustion chamber, high-pressure turbine, and low-pressure turbine.

The DSM also showed that the remaining two, more spatially distributed, subsystems were more functionally integrative across the engine. These distributed subsystems are the mechanical components and the externals and controls. They tended to have more interfaces among components of different subsystems and relatively few interfaces within each subsystem. See Sosa et al. (2003) for details and statistical tests of this analysis.

Identifying the pattern of component interfaces both within and across subsystems helped the engineering managers at Pratt & Whitney to better manage the highly complex challenge of system engineering. Their system engineering practice had been largely focused on the interactions inside the modular subsystems. Based on this analysis, they were able to focus more attention on the component interfaces across the subsystems.

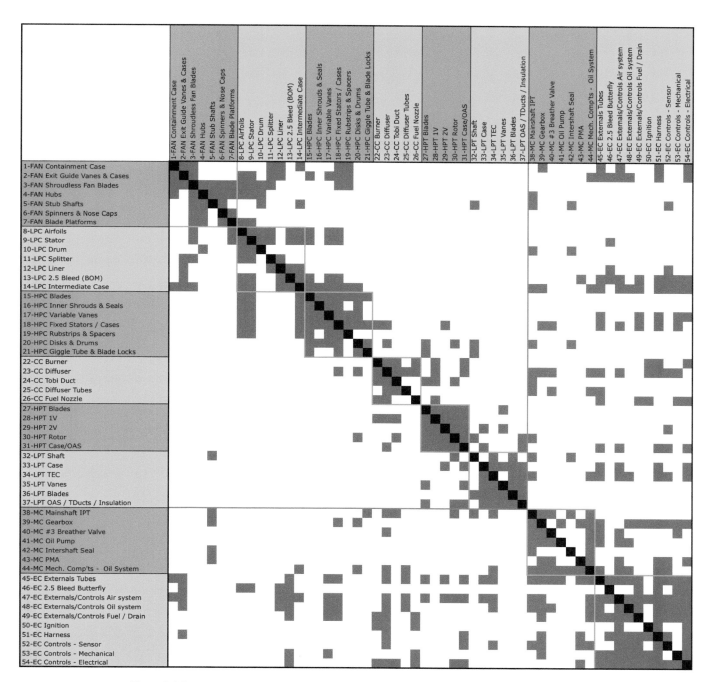

Figure 3.2.2
PW4098 Jet Engine System Architecture DSM.

References

Rowles, Craig M. 1999, February. *System Integration Analysis of a Large Commercial Aircraft Engine*. Master's thesis, Massachusetts Institute of Technology, Cambridge, MA.

Sosa, Manuel E. 2000, June. *Analyzing the Effects of Product Architecture on Technical Communication in Product Development Organizations*. PhD dissertation, Massachusetts Institute of Technology, Cambridge, MA.

Sosa, Manuel E., Steven D. Eppinger, and Craig M. Rowles. 2003. Identifying Modular and Integrative Systems and Their Impact on Design Team Interactions. *Journal of Mechanical Design* 125 (2):240–252.

Sosa, Manuel E., Steven D. Eppinger, and Craig M. Rowles. 2007. A Network Approach to Define Modularity of Components in Complex Products. *Journal of Mechanical Design* 129 (11):1118–1129.

Example 3.3   Xerox Digital Printing Technology Infusion

Contributors

Eun Suk Suh
Xerox Corporation

Olivier de Weck
Massachusetts Institute of Technology

Problem Statement

Xerox is a leading designer and manufacturer of digital printing systems such as the one shown in figure 3.3.1. These printing presses can produce several million high-quality color publications per month, including yearly corporate reports, books, marketing brochures, and other documents that are subject to stringent media and image quality requirements. The market for these machines is growing steadily—mainly at the expense of traditional offset printing—and it is also highly competitive. Firms compete with features such as versatility, print quality, system availability, and cost per print, as well as ancillary service offerings. They continuously innovate and infuse a stream of new features and technologies into their machines. The main problem addressed in this example is assessment of the potential performance benefits and the invasiveness of proposed new value-enhancing technologies into a baseline product architecture. The specific technology considered is an image density correction subsystem that automatically senses imperfections on the photoreceptor belt and digitally inverts these undesired features in the digital input to

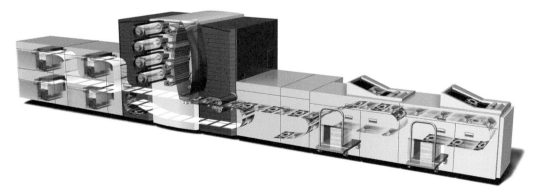

Figure 3.3.1
Xerox iGen3 Digital Printing System (courtesy of Xerox Corp.).

achieve both higher image quality and lower operating costs. A DSM was created to capture the baseline product architecture, while a so-called delta-DSM (ΔDSM) is used to document all the required component and interface changes.

Data Collection

Over the course of several months in 2007, a full DSM model of the iGen3 digital printing system was created by Eun Suk Suh (currently a system architect at Xerox and previously a doctoral student in the MIT Engineering Systems Division). The first decision that had to be made was the level of decomposition or abstraction at which the machine should be represented. If the iGen3 digital printing system, shown in figure 3.3.1, were to be completely decomposed to individual part numbers, it would result in a DSM with about 2,000 rows and columns. Because this would have been impractical, we decided to represent the system using a DSM of 84 components, as this size balanced the requirements of data collection with the desire to obtain a detailed view of the product architecture. A total of 140 person hours were spent in creating the DSM model. This included reading assembly drawings and schematics, physically inspecting a prototype machine, and interviewing experts to verify that all important components and interfaces had been properly captured. Based on this experience, the effort for manually creating a product architecture DSM scales approximately with $T = 0.02*N^2$, where T is the number of person hours and N is the number of components (parts or subsystems) represented in the DSM model.

Model

The product architecture DSM of the iGen3 digital printing system was created using the DSM template shown in figure 3.3.2. Each of the four types of interfaces—physical

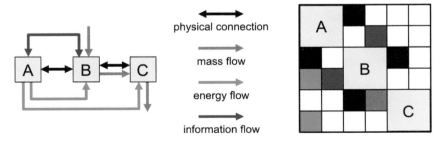

Figure 3.3.2
Block diagram (left) and corresponding DSM (right) of a simple system. Each DSM cell is subdivided to represent four types of interfaces (black = physical connection, red = mass flow, green = energy flow, blue = information flow).

connections, mass flows (e.g., toner, paper), energy flows (mechanical rotary, electrical, thermal), and information flows (image data, sensor signals, actuator commands)—was identified and included as subcells in the DSM (figure 3.3.3). This was important because the amount of effort in redesigning each of these components and types of interfaces in the new product is quite different. Figure 3.3.2 shows how to read a highly simplified DSM of this type for a simple system composed of three components A, B, and C. Following this template, the full iGen3 digital printing system DSM is shown in figure 3.3.3.

Some noteworthy measures of complexity include the fact that there are 572 physical connections (black), 45 different mass flows (red), 167 energy flows (green), and 165 information flows (blue) in the system. While complex, the density of the system is only 3.7%. In other words, only 1,033 of the 27,972 off-diagonal cells are occupied. Part of the reason that the effort for creating such a model scales with N^2 and not N is that the empty cells also need to be confirmed.

We next built a ΔDSM from the DSM. The ΔDSM is based on the underlying product architecture DSM but captures only the engineering changes that are required to add to the system a new set of components related to a proposed new technology. The following steps were taken to construct the ΔDSM:

1. Empty all cells of the baseline DSM (figure 3.3.3).
2. To the baseline DSM, add new rows and columns for any newly added components and insert the names of the new components.
3. For newly added, removed, or modified components and connections, fill in the corresponding cells of the ΔDSM using the color coding scheme shown in figure 3.3.2.
4. Note that both changes directly required by the new technology as well as indirect (propagated) changes should be included in the ΔDSM.

Using these guidelines, a ΔDSM for the new technology was constructed. Figure 3.3.4 shows the completed ΔDSM for the new technology. In the figure, only those elements that are affected by technology infusion are shown (rows and columns without any change are deleted). Overall, 15 components were added, eliminated, or revised. There were 33 physical connection changes, no mass flow changes, 7 energy flow changes, and 32 information flow changes, for a total of 87 changes in the system.

Results

The ΔDSM is used to assess the anticipated effort for designing and infusing the new technology into the baseline product. This can be done in two ways. First, one can simply assess what fraction of the original product is affected by the new technology. This fraction is referred to as the Technology Invasiveness Index (TII) and is computed as follows:

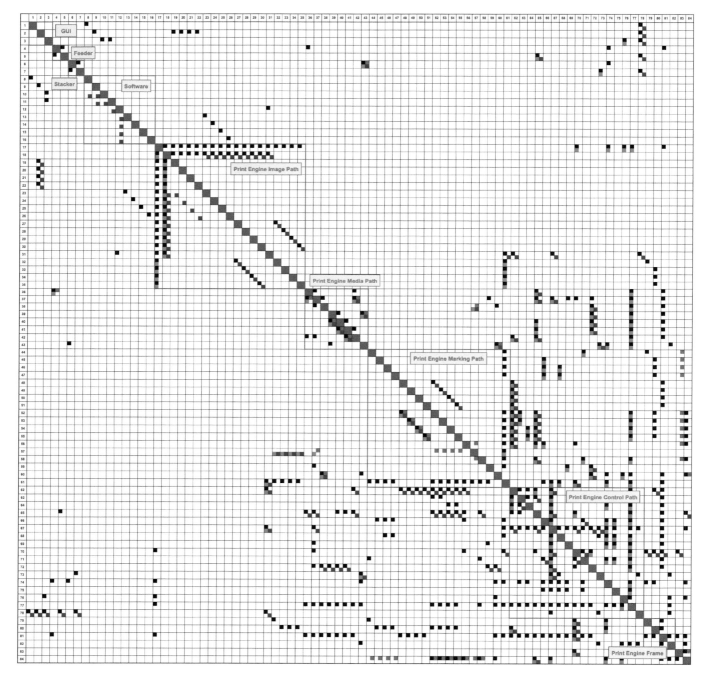

Figure 3.3.3
Product architecture DSM of the Xerox iGen3 digital printing system, indicating four types of interfaces across the 84 components, grouped into nine subsystems.

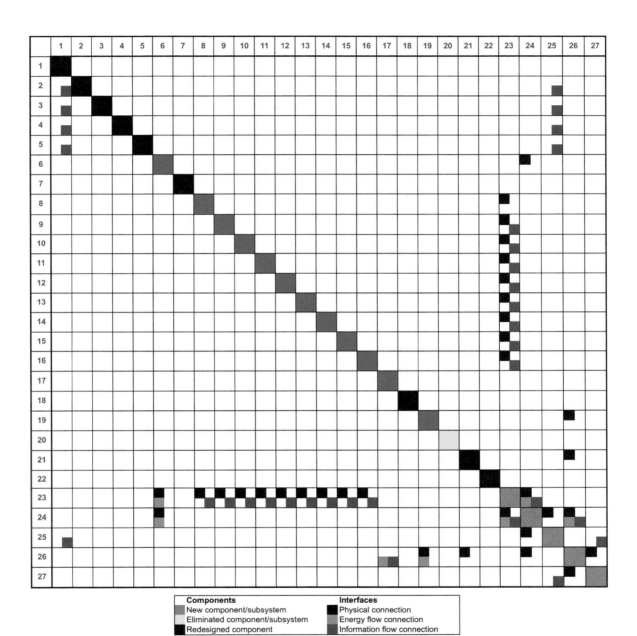

Figure 3.3.4
Delta-DSM for auto-density image correction technology.

$$TII = \frac{NEC_{\Delta DSM}}{NEC_{DSM}} = \frac{\sum_{i=1}^{N2}\sum_{j=1}^{N2}\Delta DSM_{ij}}{\sum_{i=1}^{N1}\sum_{j=1}^{N1}DSM_{ij}}$$

where

$NEC_{\Delta DSM}$ is the number of non-empty cells in the ΔDSM (figure 3.3.4)

NEC_{DSM} is the number of non-empty cells in the baseline DSM (figure 3.3.3)

$N1$ is the number of elements in the baseline DSM

$N2$ is the number of elements in the ΔDSM

TII represents the relative system change magnitude with respect to the complexity of the original system due to technology infusion. For the technology examined here, TII was calculated to be 8.5%. The changes relate to the physical integration of additional sensors as well as changes in the electrical power and control subsystems.

A second way to assess impact from the ΔDSM is to estimate the amount of resources and effort needed to make each individual design change and the effort associated with system integration and testing. Two changes may contribute equally to TII but may require vastly different amounts of resources to implement. Usually, experts from relevant fields are consulted to estimate the amount of engineering effort and investment required to accommodate changes specified in the ΔDSM. This is then translated into cost. In the case of the image correction technology presented here, the total effort was estimated to be 13 person years. This is the required up-front investment for infusing the technology into the product. The analysis of effort required needs to be complemented along with an analysis of the performance impact on the attributes that customers value, and that may lead to additional sales and expected profit. In this particular example, Xerox decided to include the new technology as part of the next-generation iGen4 digital printing system, which was launched in 2008 and received several awards for its high level of performance.

References

de Weck, O. L. 2007, October 16–18. *On the Role of DSM in Designing Systems and Products for Changeability.* Ninth International Design Structure Matrix Conference, DSM'07, Munich, Germany.

Suh, E. S., O. L. de Weck, and D. Chang. 2007. Flexible Product Platforms: Framework and Case Study. *Research in Engineering Design* 18 (2):67–89.

Suh, E. S., M. R. Furst, K. J. Mihalyov, and O. de Weck. (2010, Summer). Technology Infusion for Complex Systems: A Framework and Case Study. *Systems Engineering* 13 (2):186–203.

Example 3.4   NASA Mars Pathfinder Technology Readiness

Contributors

Tim Brady
NASA

Deborah Nightingale
Massachusetts Institute of Technology

Problem Statement

The U.S. National Aeronautics and Space Administration (NASA) has a broad mission—to conduct human and robotic space exploration, scientific discovery, and aeronautics research. In the mid- to late 1990s, NASA launched several robotic spacecraft missions to demonstrate new technology while also executing these missions with shorter development times. Successes in this approach included the landing of the Mars Pathfinder in 1997, which provided close-up views of the Martian surface and demonstrated the use of a small, robotic rover (figure 3.4.1). The successes were offset with some failures, most notably the loss of both the Mars Climate Orbiter and Mars Polar Lander in 1999. These failures motivated investigation of the effectiveness of DSM to provide a comprehensive system view of the product architecture and the effect of technology maturity and risk in system components.

Data Collection

Over the course of five months in 2001, Tim Brady (both a NASA employee and a student in MIT's System Design and Management master's program) researched seven robotic spacecraft missions, six led by NASA and one led by the Department of Defense. The cases selected had complex missions with budgets ranging from $30 to $300 million and development times of approximately three years. Cases were also selected based on availability of data related to the spacecraft architecture and subsystem technology maturity. One of these cases was the Mars Pathfinder spacecraft, which landed successfully on the surface of Mars in July 1997 and deployed a robotic rover.

Model

A technology risk DSM (TR-DSM) is based on a product architecture DSM using a decomposition of the major components of the spacecraft. The TR-DSM is generated using a three-step process. In the first step, a product architecture DSM is generated using values for the strength of each component interface dependence.

Figure 3.4.1
The Mars Pathfinder rover in a simulated test environment (courtesy of NASA).

The interface dependency value assigned to the DSM cell is obtained by summing values representing the physical, energy, and information interactions that exist between a pair of elements. In this DSM example, a physical interface value of 2 is assigned where a direct physical interface exists. An energy interface value of 2 is assigned where there is direct energy transfer such as power, propulsion, or thermal loads. The information interface was assigned a value of 2 where there is direct transfer of information between components and a value of 1 where information is transferred indirectly between components.

In the second step of the TR-DSM generation, each component is assigned a technology risk factor (TRF). The TRF scale ranges from a value of 1 for the most mature components to a value of 5 for the highest risk or unproven components. The specific value assigned is based on criteria set by NASA's technology readiness level (TRL) definitions shown in figure 3.4.2. In the TR-DSM (see figure 3.4.1), a column and row are added next to the component names, and the TRF values are placed in the DSM cell adjacent to the component name.

The final step of the TR-DSM generation is calculating the value to be placed in each cell of the DSM using the following formula:

TRF	NASA TRL Definition	TRL
1	Actual system "flight proven" through successful mission operations	9
2	Actual system completed and "flight qualified" through test and demonstration	8
2	System prototype in a space environment	7
3	System/subsystem model or prototype demonstration in a relevant environment	6
4	Component and/or breadboard validation in relevant environment	5
4	Component and/or breadboard validation in laboratory environment	4
5	Analytical and experimental critical function and /or characteristic proof-of-concept	3
5	Technology concept and/or application formulated	2
5	Basic principles observed and reported	1

Figure 3.4.2
Technology Risk Factor (TRF) and Technology Readiness Level (TRL) definitions (NASA 2008).

$$\frac{\text{TRF of}}{\text{component A}} \times \frac{\text{TRF of}}{\text{component B}} \times \frac{\text{interface dependency}}{\text{value A-to-B}} = \frac{\text{A-B interface value in}}{\text{technology risk DSM}}$$

The TR-DSM for the Mars Pathfinder spacecraft is shown in figure 3.4.3. This DSM was created to demonstrate application of TRLs to compute areas of high risk in space system development.

Results

The TR-DSM can be used to highlight areas of development and operational risk. One of the major objectives of the Pathfinder mission was to demonstrate new technologies that could help reduce the cost of delivering scientific instruments to Mars. These components included a radiation-hardened computer based on commercial hardware, utilization of distributed processors linked together with a data bus, telecommunications circuit boards, and components that supported the strategy for aero-braking entry, parachute descent, and touchdown with airbags surrounding the lander. The majority of these advanced technology systems were tested in simulated Martian environments on Earth and assigned a TRF value of 3 based on the criteria in figure 3.4.2. The aeroshell used during the entry into the Martian atmosphere could not be fully tested on Earth and was assigned a TRF value of 4.

The resulting TR-DSM shown in figure 3.4.3 identifies several clusters of technology risk areas. For example, the entry, descent, and landing (EDL) subsystem shows up as an area of high technology risk. The high values result from a set of interfaces identified with relatively high dependence between components with high technology risk factors. The interfaces associated with the telecommunications, the landing instrumentation, and the rover also showed clusters of high technology risk.

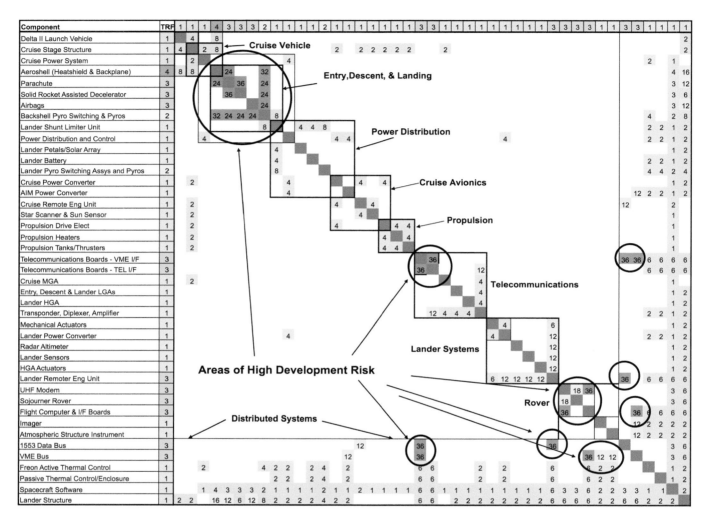

Figure 3.4.3
Technology Risk DSM for Mars Pathfinder.

The Mars Pathfinder project had an exceptional risk management approach, and the case study can be used to assess the effectiveness of the TR-DSM in identifying the same project risks. The largest pattern of high-risk numbers in the TR-DSM was consistent with observations made by the project manager. Following an early investigation into the nature and potential for development risk in each subsystem, project manager Anthony Spear noted, "To no one's surprise, the EDL phase emerged as the biggest Mission risk, with the airbags as the most risky EDL element."

The TR-DSM can be used as an analytical tool throughout a project's development life cycle for identifying and communicating high-risk areas in a single-system view. High TRF values can be used to identify subsystems and components requiring a thorough mitigation strategy during development.

References

Brady, Timothy K. 2002. *Utilization of Dependency Structure Matrix Analysis to Assess Complex Project Designs.* Proceedings of ASME Design Engineering Technical Conferences, no. DETC2002/DTM-34031, Montreal, Canada.

NASA Research and Technology Program and Project Management Requirements. 2008. Appendix J: Technology Readiness Levels, NASA Procedural Requirement, no. NPR 7120.8.

Spear, Anthony J. 1999. Mars Pathfinder's Lessons Learned from the Mars Pathfinder Project Manager's Perspective and the Future Road. *Acta Astronautica* 45 (4–9):235–247.

Example 3.5   Mozilla Software Redesign Effort

Contributors

Alan MacCormack, Carliss Baldwin, and John Rusnak
Harvard Business School

Problem Statement

Many firms experience significant costs related to maintaining legacy software systems and adapting these systems to uncertain future demands. These costs can be reduced by "refactoring" efforts—changes to the design that have the impact of reducing system complexity while maintaining overall system functionality. Unfortunately, we lack robust methods and metrics for evaluating the impact of these redesign efforts.

In this work, we applied DSM-based methods to explore the impact of a single major software redesign effort (MacCormack et al. 2006). We focused on the Mozilla web browser, a product derived from a commercial web browser called Navigator, which was developed by Netscape. Mozilla was released as open source code in early 1998, with the hope that volunteer developers would contribute to its ongoing development. Shortly thereafter, however, it became clear that it was difficult to contribute to Mozilla given the level of interdependency between the system's components. Hence, a small team of developers decided to redesign the system, with the intention of making the code more modular and, hence, easier to work with. We examined the design of Mozilla before and after this redesign effort.

Data Collection

The source code for Mozilla is hosted online and is freely available to everyone because it is distributed as open source code. We accessed all versions of Mozilla that were released in 1998. We processed the source code of each version through a static analysis tool called *Understand C* (distributed by Scientific Toolworks) to identify the dependencies between source files. We focused on one important dependency type—the "function call"—identified in prior work as an important determinant of modular structure (Banker et al. 2000; Rusovan et al. 2005). Function calls are requests by one part of the system to execute functionality contained in another. We generated a DSM by plotting function call dependencies between source files organized by the directory structure of the system (i.e., a nested hierarchy of modules arranged alphabetically within layers). We constrained the DSM to contain only binary values given that the distribution of function calls between system elements was highly skewed. We chose to focus only on C files in our analysis, excluding header files, which are much smaller in size and play a different role with regard to system function.

To understand the level of coupling in a system, we computed the level of visibility (reachability) for each component. (Chapter 6 illustrates a way to perform this computation.) Visibility captures all of the direct *and indirect* dependencies that a component has with other components. Because function calls have directionality, visibility is not a symmetric measure—fan-in visibility (function calls received) and fan-out visibility (function calls made) may differ for a given component. The mean level of visibility for a system, however, will be identical in each direction (i.e., each outgoing call will have a corresponding incoming call). The visibility matrix for a system is computed by calculating the transitive closure of the first-order dependency matrix. The density of this matrix is called the system's *propagation cost*. Intuitively, this metric captures the proportion of a system's elements that could be affected, on average, when a change is made to one randomly chosen element.

Model

Figure 3.5.1 shows the DSMs from two releases of the Mozilla web browser. The left side shows the DSM for a version of the software before the redesign effort. The right side shows the DSM for the version of the software immediately after the redesign effort.

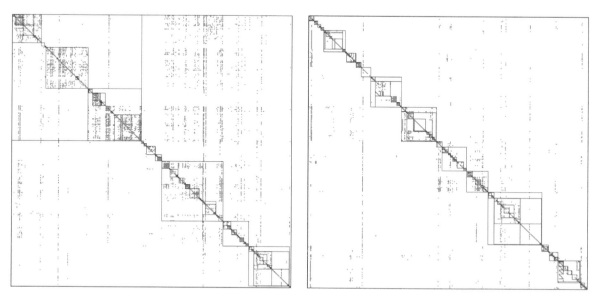

Figure 3.5.1
Mozilla software architecture DSMs, before (left) and after (right) the redesign effort.

Results

The contrast between the two designs is striking, both visually and quantitatively. The redesigned version of Mozilla consists of smaller modules (directories) with fewer dependencies between them. The system has a significantly lower dependency density—0.13% versus 0.24%. In addition, the propagation cost has declined dramatically, from 17.35% to 2.78%. In summary, the redesign had the effect of lowering the potential impact of changes to the system design by more than 80%.

It is insightful to look at the impact of this redesign effort in the context of the evolution of Mozilla's design over time. To this effect, we plot the evolution of Mozilla's propagation cost for subsequent releases in figure 3.5.2. The results once again highlight the value of this type of analysis. Prior to the redesign, Mozilla's level of coupling varied between 15% and 18%. After the redesign, Mozilla's level of coupling consistently fell to between 2% and 6%. We conclude that the redesign effort had a significant and sustained impact on reducing the cost of changes to this system.

Our work demonstrates that the application of DSM-based methods can help reveal the impact of architectural redesign efforts on complex systems. In this case, a small, focused team of developers achieved substantial reductions in system complexity over a period of less than four months. These improvements substantially reduced the effort required to contribute to the Mozilla project, given each component was coupled to fewer other components. Contributors needed to understand less of the code to make a

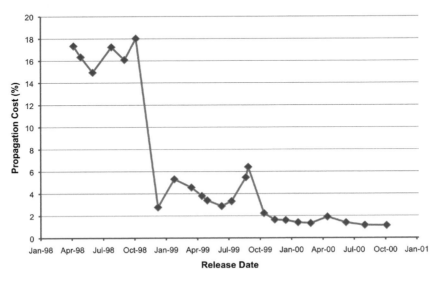

Figure 3.5.2
Evolution of Mozilla's propagation cost over time.

contribution. Each of the changes they did make had a lower probability of affecting other functions in a negative way. Subsequently, the Mozilla project developed a vibrant community of programmers willing to contribute new code. The Mozilla software became the foundation for the highly successful Firefox Internet browser.

In related work, we have used similar methods to tackle a variety of important questions that require the use of robust and repeatable measures of system architecture. We have shown that successful open source projects, in general, generate products with more modular architectures than the equivalent commercial software (MacCormack et al. 2011). We have found that measures of component visibility predict design evolution in terms of component survival, augmentation, and change (MacCormack 2009). Finally, we have shown how measures of visibility can be used to characterize different types of systems, thereby revealing the degree to which each has a "core-periphery" structure (MacCormack et al. 2010).

References

Banker, Rajiv D., and Sandra A. Slaughter. 2000. The Moderating Effect of Structure on Volatility and Complexity in Software Enhancement. *Information Systems Research* 11 (3):219–240.

MacCormack, Alan. 2009. *The Impact of Component Modularity on Design Evolution: Evidence from the Software Industry*. Harvard Business School, Working Paper no. 08–038.

MacCormack, Alan, John Rusnak, and Carliss Baldwin. 2006. Exploring the Structure of Complex Software Designs: An Empirical Study of Open Source and Proprietary Code. *Management Science* 52 (7):1015–1030.

MacCormack, Alan, John Rusnak, and Carliss Baldwin. 2010. *The Architecture of Complex Systems: Do Core-Periphery Systems Dominate?* Harvard Business School, Working Paper no. 10–059.

MacCormack, Alan, John Rusnak, and Carliss Baldwin. 2011. *Exploring the Duality between Product and Organizational Architectures: A Test of the Mirroring Hypothesis*. Harvard Business School, Working Paper no. 08–039.

Rusovan, Srdjan, Mark Lawford, and David Lorge Parnas. 2005. Open Source Software Development: Future or Fad? In *Perspectives on Free and Open Source Software*, ed. Joseph Feller et al. Cambridge, MA: MIT Press.

Example 3.6   AgustaWestland Helicopter Change Propagation

Contributors

John Clarkson, Caroline Simons, and Claudia Eckert
Engineering Design Centre, University of Cambridge

Problem Statement

AgustaWestland produces helicopters for civil and military applications, such as the AW101 aircraft shown in figure 3.6.1, utilizing an array of world-leading technologies. In providing products for particular customers, aircraft designs are often based on an existing model but redesigned or customized to meet specific needs. During this process, a change to one part of the product will in most cases result in changes to other parts. The prediction of such change provides a significant challenge in the management of redesign and the customization of complex products where many change propagation paths may be possible. This DSM application demonstrates a model to predict the risk of change propagation in complex products.

Figure 3.6.1
A military version of the AW101 helicopter (courtesy of João Paulo Nabais).

Data Collection

Interviews were conducted in 1999 with 17 senior engineers working on various aspects of the helicopter design (such as stress analysis, load modeling, or fuselage design) to assess the scope and complexity of change in existing product ranges. A further four interviews with chief engineers or deputy chief engineers and an interview with a manager responsible for producing proposals for new projects focused on understanding the changes involved in generating a new version of a helicopter. Subsequently, a meeting was held with seven senior engineers to discuss change propagation and agree on the structure of the binary DSM for the AW101. Then, two attributes of each interface—likelihood and impact of change from one component to the other—were elicited from deputy chief engineers by AgustaWestland staff for use in the propagation analysis. Finally, details of a number of redesign cases were obtained to provide clear evidence of change propagation and to assist validation of the analysis method.

Model

The product architecture DSM model shown in figure 3.6.2 comprises 19 key components and subsystems, all based on the same product architecture—a simplified description of the helicopter. The original data on likelihood and impact of change between adjacent components are stored in a conventional product architecture DSM. The derived likelihood and impact of change, taking account of all possible change propagation paths, is also stored in another DSM. Details of the analytical method for computing the change propagation results are presented in several research publications from Cambridge EDC (cited below).

The change propagation DSM shown in figure 3.6.3 represents the combined risk of changes propagating between systems, both directly and indirectly, with the columns having impact on the rows (IC convention). The width of the rectangle in each cell depicts the likelihood of change (certainty represented by the full width of the cell), and the height depicts the impact of change (complete redesign represented by the full height of the cell). Thus, with risk calculated as the product of likelihood and impact, the size of the shaded area of each off-diagonal cell in the DSM conveys the amount of risk. Red shading signifies a significant risk of change propagation, amber a lower risk, and green a small risk of propagation. Interestingly, almost every off-diagonal cell is filled.

Results

The nature and extent of change propagation is generally neither clearly understood nor well predicted. However, it can cause large delays or unexpected spending in design projects. A change analysis method was developed using a product DSM to assist in the

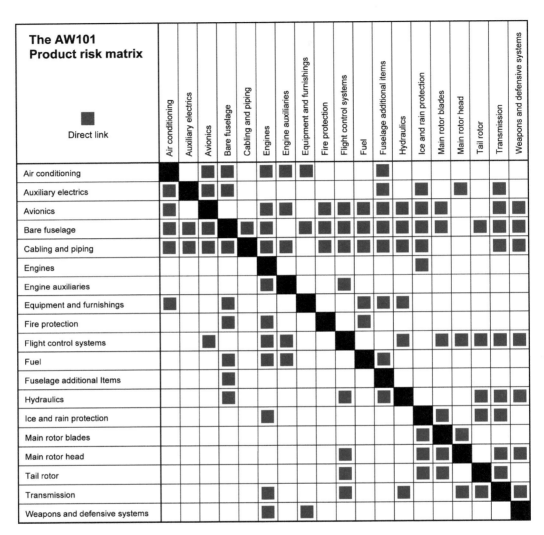

Figure 3.6.2
AW101 product architecture DSM.

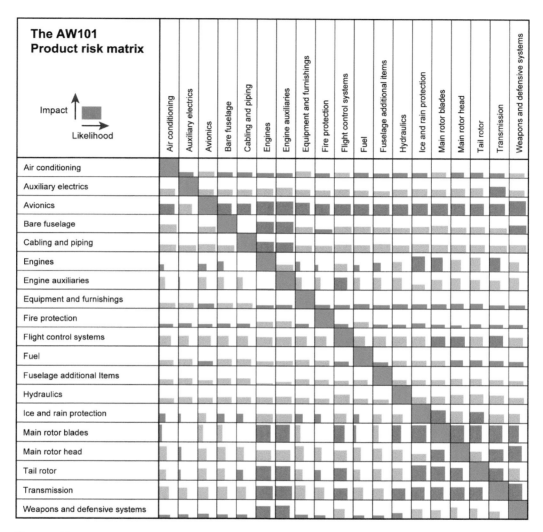

Figure 3.6.3
AW101 change propagation DSM.

prediction of change propagation in complex products. It appears to provide useful insight into the change behavior of complex systems, such as a helicopter. In the AW101 case, the most significant changes are propagated from many different components and systems to several others, such as avionics and main rotor blades. Many such change paths shown in the change propagation DSM are not initially identified in the product architecture DSM, which represents only direct interactions between components. Propagated changes are predicted by the change method and are also documented to have happened in practice at AgustaWestland.

Experience from a number of additional case studies has shown that the time taken to build a moderately sized model (fewer than 50 components) is acceptable, and the companies involved all found the process valuable. Clearly more work remains to be done. The analytical method used relies on many assumptions, the validity of which need to be further explored. However, the need for and the possible success of the change prediction method seems clear.

References

Clarkson, P. John, Caroline S. Simons, and Claudia M. Eckert. 2004. Predicting Change Propagation in Complex Design. *Journal of Mechanical Design* 126 (5):765–797.

Eckert, Claudia M., P. John Clarkson, and Winfried Zanker. 2004. Change and Customisation in Complex Engineering Domains. *Research in Engineering Design* 15 (1):1–21.

Jarratt, Tim A. W., and Claudia M. Eckert, Nicholas H. M. Caldwell, and P. John Clarkson. 2011. Engineering Change: An Overview and Perspective on the Literature. *Research in Engineering Design* 22 (2):103–124.

Example 3.7   Johnson & Johnson Clinical Chemistry Analyzer

Contributor

Qi D. Van Eikema Hommes
Massachusetts Institute of Technology

Problem Statement

Ortho-Clinical Diagnostics (OCD) is a medical device company within Johnson & Johnson. We studied OCD's OASIS clinical chemistry analyzer—a system typically used in large hospitals to automate the testing of patients' blood and other body fluids. The analyzer is a complex system containing electromechanical systems, software, as well as wet and dry chemistry. The size of the analyzer is similar to a large office copy machine. At the peak of the development process in 2001, the core development group had approximately 120 engineers and scientists.

The OASIS analyzer (named VITROS on the market, one model of which is shown in figure 3.7.1) was the first analyzer OCD designed to incorporate wet chemistry. Previous OCD products only had thin-film technology. Wet chemistry technology, however, has been applied for many years in competitors' products. Therefore, the design challenge was not the technology but rather the integration of two mature technologies into a new product that was more complex than previous products.

Figure 3.7.1
A Clinical Chemistry Analyzer (courtesy of Ortho-Clinical Diagnostics, Inc.).

When the case study started, the OASIS program was in the early detailed design phase. After seeing our presentation introducing the DSM method, the OCD engineers wanted to build a product architecture DSM in order to capture their understanding of the interactions among the subsystems in the analyzer. They believed that the system interaction knowledge captured by the DSM would help the system engineers anticipate potential system integration issues and in turn prevent design rework and schedule delays late in the program.

In research at MIT, we had developed a matrix transformation method to predict system interactions based on product requirements without relying on the experts' knowledge about the detailed design (Dong 2002; Dong and Whitney 2001). This method starts with a Design Matrix (DM) mapping system design parameters to functional requirements of the system. (This is a type of domain mapping matrix [DMM], which is discussed further in chapter 8.) By selecting the diagonal elements of the DM as the output variables, the DM can be turned into a DSM, representing the interactions among the design parameters in the system. Using the DM-DSM matrix transformation technique, engineers can predict the interactions between components in the system based on how they work together to fulfill the functional requirements.

Because the medical device industry is highly regulated, the product design requirements were well documented. Therefore, it was possible to compare the product architecture DSM constructed by the engineering experts with a DSM from the matrix transformation method in hopes of maximizing our understanding of the system interfaces and minimizing system integration risks.

Therefore, the objectives for this case study were to:

1. Build a product architecture DSM to capture system interactions based on experts' knowledge of the product design during the detailed design phase.

2. Build a product architecture DSM using product requirements and the matrix transformation method to predict system interactions that exist in the designed analyzer.

3. Compare and combine the results in 1 and 2 to obtain a comprehensive prediction of the system interfaces in order to assist the system integration efforts of the OASIS analyzer.

Data Collection

We focused this case study on the interactions among the major subsystems of the OASIS analyzer for two reasons. First, this was the level of detail at which the systems engineering team was working. Second, the amount of design details at this level was sufficient to provide insights into the system but not too much for a three-month, one-person project.

The OCD engineers built two DSMs, one in February 2001 and one in August 2001, documenting their progressive understanding of the system as more detailed design decisions were made. I spent three summer months at OCD as a researcher to construct the prediction DSM based on requirements without knowledge of the DSMs that the OCD engineers built. The OCD engineers and scientists served as consultants when I had questions regarding the product and technology.

Model

The DSM model shown in figure 3.7.2 is the Expert DSM, representing the combined results of two DSM building exercises led by the OCD engineers and scientists in

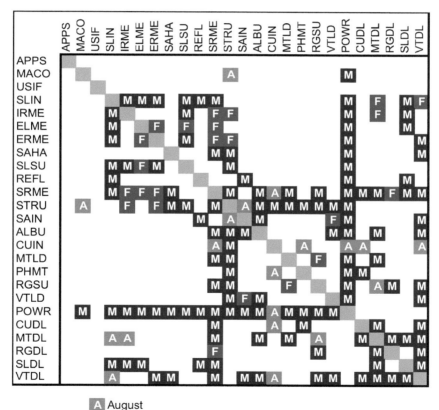

Figure 3.7.2
Expert DSM, a consolidation of two DSMs produced by engineering experts.

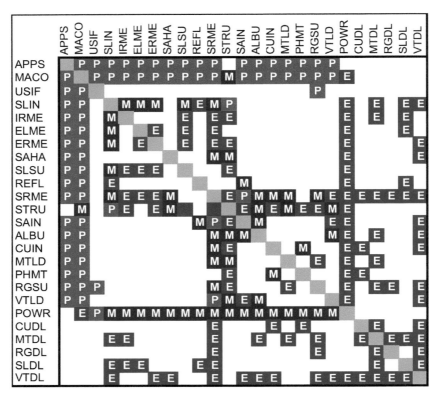

P Prediction DSM
E Expert DSM
M Both in the Prediction DSM and in the Expert DSM

Figure 3.7.3
Comparing the Prediction and Expert DSMs.

February and August 2001. The row and column headings are abbreviations for the major subsystems in the analyzer. The two DSM building exercises did not produce identical results. The overlaps and differences are identified using symbols and colors in the DSM.

I built a Prediction DSM from design requirements using the matrix transformation method and compared to the Expert DSM. Figure 3.7.3 shows the results of this comparison. Symbols and colors in the DSM identify the overlaps and differences.

Results

This project produced two important insights: (1) the completeness of the DSM depends on the coverage of topics during the DSM building exercises; and (2) the matrix transformation method using product requirements can predict many system interactions that will happen later in the design process, including those that expert engineers may miss using the traditional DSM construction approach. These two results are discussed in more detail below.

The DSMs produced by the engineers in February and August (figure 3.7.2) display evolution of their design knowledge during the project. The February DSM did not contain the interactions marked "A" because the design matured over the course of six months. However, the DSM construction exercise in August missed the interactions marked "F." Engineers reviewed the February interactions in August and admitted that they were still valid but had been missed in August because no one remembered to talk about reliability issues during that DSM construction exercise.

In addition, figure 3.7.3 shows that the Prediction DSM based on design requirements captured system interactions that engineering experts did not capture in the two DSMs that they built (labeled "P" in figure 3.7.3). The engineers missed most of these interactions because they did not invite the software engineers to the DSM building exercises. Historically, OCD products were mostly electromechanical systems. The system engineers did not realize how intertwined the software system actually was with the rest of the hardware system. Without the Prediction DSM, there could have been unanticipated system interactions causing delays and rework late in the system integration phase. Hence, one of the lessons learned from this case study is that the quality of the DSM constructed depends heavily on who was invited to provide inputs to the DSM. Any DSM construction exercise must identify key stakeholders of the system and all of the important design issues that need to be considered.

The Prediction DSM based on system design requirements missed many marks in the Expert DSM because the engineers did not give the author the assay chemistry requirements. Therefore, the system interactions related to assay chemistry were not predicted. If I had been given the assay requirements documents, then 76% of all of the marks in figure 3.7.3 could have been predicted by the requirements-based matrix transformation method. In addition, the matrix transformation method also predicted another 3% more interactions that engineering experts missed in their discussions. Therefore, the requirements-based matrix transformation method, if used in the early stage of the design process when requirements are understood but detailed design is not yet available, can be a very powerful technique to anticipate areas in the system that may cause integration issues and rework. Such understanding early on in the product development process will make system integration efforts less reactive. This technique can also help engineers to compare and choose system design concepts that minimize system integration risks.

At the end of this project, we combined all of the system interfaces learned from the three DSM construction exercises to form the DSM in figure 3.7.3. This DSM offers a comprehensive view of the system interactions. The DSM model produced many insights for the OCD engineers about their system design. OCD system engineers used it to guide the system design and integration efforts for the OASIS analyzer.

References

Dong, Qi. 2002. Predicting and Managing System Interactions at Early Phase of the Product Development Process. PhD thesis (Mechanical Engineering), Massachusetts Institute of Technology, Cambridge, MA.

Dong, Qi, and Daniel E. Whitney. 2001, September 9–12. *Designing a Requirement Driven Product Development Process*. DTM21682, Proceedings of ASME Design Engineering Technical Conferences, Pittsburgh, PA.

Dong, Qi, and Daniel E. Whitney. 2003, September 2–6. *The Predictability of System Interactions at the Early Phase of Product Development Process*. DETC2003/DTM-48635, Proceedings of the 2003 ASME Design Engineering Technical Conferences, Chicago, IL.

C. low dependencies inside of the layer and high dependencies outside

D. low dependencies in and outside of the layer

This characterization allows a logical expression of how components cluster, while accurately reflecting the behavior of some components provides hints about the proper placement of more sporadic components. Type *A* components tend to make up the core of their layer, while a few move outside to form system-integrating components. Type *B* components also remained inside and at the core of their layers, while many type *C* components moved to join a different layer or suggest a new module. The type *C* components that remained in the layer tended to be on the periphery of the layer. Type *D* components tended to move either to the periphery of their layer or outside their layer as isolated components or newly formed layers.

The work has led to a method to quickly characterize components, providing refined guidance for identifying components that require further design, enabling alternative modules (layers), and suggesting changes in component designs. Early involvement in the process allows a range of solutions to be visualized by the designer, helping them consider how the building's components interact and thereby negotiating more informed tradeoffs. At each design stage, an analysis of the DSM could be made, where observations (guided by the principles being developed) feed into the next stage of the design process to create refined modules with fewer dependencies outside their layer, hence creating a more adaptable solution.

References

Brand, Stewart. 1994. *How Buildings Learn: What Happens After They're Built*. New York: Penguin.

Schmidt III, Robert, Toru Eguchi, Simon Austin, and Alistair Gibb. 2009a, October 5–9. *Adaptable Futures: A 21st Century Challenge*. Changing Roles—New Roles, New Challenges, Rotterdam, the Netherlands.

Schmidt III, Robert, Simon Austin, and David Brown. 2009b, October 12–13. *Designing Adaptable Buildings*. Proceedings of the 11th International DSM Conference, Greenville, SC.

Example 3.9   Kodak Single-Use Camera

Contributors

Fabrice Alizon
Keyplatform Company

Steven B. Shooter
Bucknell University

Problem Statement

Kodak, a manufacturer of photographic equipment and systems, successfully led the market of single-use cameras by producing a product family that addressed multiple market segments. Kodak offered a wide range of products that included combinations of key features such as waterproof, panoramic format, flash, and high definition. Product platforming enables companies to cut costs while offering tailored products, yet it also brings the challenge of managing variety within the family. This DSM application demonstrates two DSM techniques to identify modules across a product family: the DSM variety (DSM^V) and the three-dimensional DSM (DSM^{3D}). Using these two DSM techniques, we are able to study families of products, modules, and interfaces.

Data Collection

We dissected five Kodak single-use cameras, including the Fun Saver model shown in figure 3.9.1. Each time a new component was identified, it received a new bill-of-material

Figure 3.9.1
Kodak Fun Saver, one of the single-use cameras studied (courtesy of Eastman Kodak Company).

(BOM) reference number. We then documented the interactions among components. By comparing the component interactions across the five camera models, we categorized each interface as common (occurring in all five), variant (occurring in some of the five), or unique (occurring in one of the five).

Model

The model works in two main stages using two original DSM techniques: DSM^V and DSM^{3D}. The DSM^V, shown in figure 3.9.2, uses a static, binary, product architecture DSM

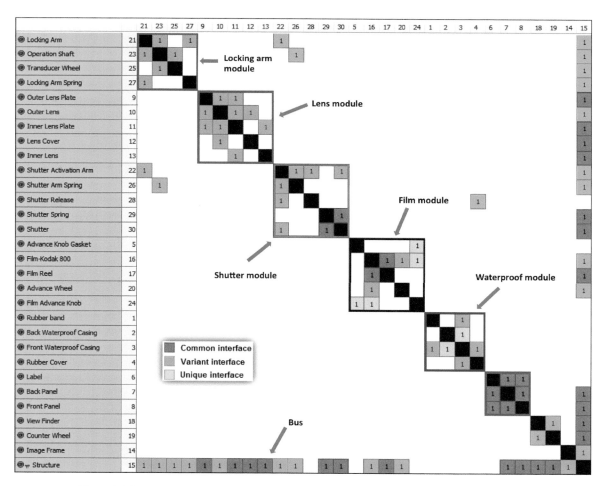

Figure 3.9.2
Clustered DSM^V for one of the cameras.

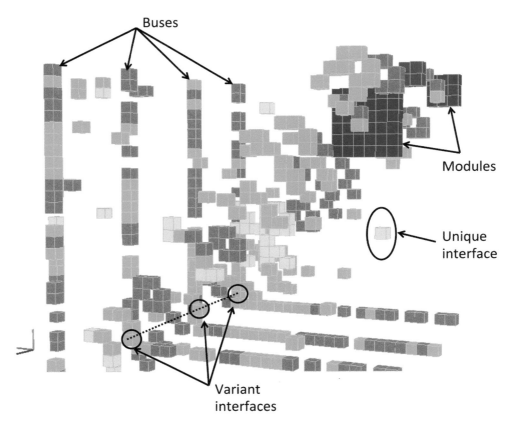

Figure 3.9.3
DSM^{3D} showing several single-use camera DSMs overlapping in 3D.

to specify the modules in each product containing components that have either common, variant, or unique interfaces. We proceed the same way for all the products of the family (total of five), and we then stack these five DSM^{V}s to obtain the DSM^{3D}. The DSM^{3D}, shown in figure 3.9.3, is a three-dimensional DSM gathering all products of the family and highlighting the differences.

Based on the interactions among components, a DSM clustering algorithm identified five modules indicated by square borders in the DSM^V. The last component in the list (Structure) is related to many other components, as indicated with the many colored squares along the bottom of the DSM. This bus-type component is strategic because there is an opportunity to use common interfaces to save cost and better handle the diversity. We see in the DSM^V that many interfaces are variant.

The DSM^V enables designers to study basic aspects of product architecture such as bus, mini-bus, and strength of physical interactions. It also helps to investigate the architectural distribution of modules and interfaces.

Results

The analysis was done through Synerg' (Alizon 2009), a software application developed to manage product family design. Each DSM was clustered using an original algorithm, which can cluster a product DSM of more than 1,000 components in only a few minutes.

Interfaces having instantiation in multiple products are named *cross-interfaces*. Modules having instantiation in multiple products are named *cross-modules*. By identifying cross-interfaces and cross-modules, we believe this was the first DSM to analyze the overall architecture of the product family. DSM^V provides the current diversity of each product, whereas the DSM^{3D} highlights the differences among these products. It is beneficial to have a red and yellow DSM^{3D} such that all elements are either common (red) or unique/specific (yellow). Designers should avoid blue, which represents variant interfaces that provide diversity and additional cost.

Cross-Interface Management

These DSM representations assist in decision making and the exploration of alternatives when developing or refining a product family. Consider a straightforward example where four interfaces are common and the fifth one is unique/specific. This single interface that is unique/specific results in a cross-interface characterized as variant. Alteration of that one interface can dramatically improve the product family. Although it is possible to identify and interpret this scenario without using DSM^{3D}, one can see the value in more complex architectures with more challenging interfaces.

This tool can help designers to:

- Try to design a new common interface handling the common and unique interfaces;
- Communicate with other services (such as cost management and marketing) to ultimately negotiate for a common component, leading to a common interface; and
- Financially justify to product management the solution using a variant cross-interface.

Cross-Module Management

When a cross-module is variant, designers can focus on this cross-module to develop a common module and common interfaces in an effort to reduce cost.

DSM^V and DSM^{3D} are combined in a single process to better manage both modularity and variety. The DSM^V models common, variant, and unique modules and interfaces across products and enables one to study these in detail. The DSM^{3D} permits a higher level of analysis for an entire family of products and for cross-modules and cross-components to study their specification and interfaces.

References

Alizon, Fabrice. 2009, February. *Module-Based Design Management-Synerg'*. Symposium on Product Family & Product Platform Design, Helsinki University of Technology (TKK), Helsinki, Finland.

Alizon, Fabrice, Seung K. Moon, Steven B. Shooter, and Timothy W. Simpson. 2007, September 4–7. *Three Dimensional Design Structure Matrix-DSM3D*. ASME Design Engineering Technical Conferences, DETC2007–34510, Las Vegas, NV, pp. 941–948.

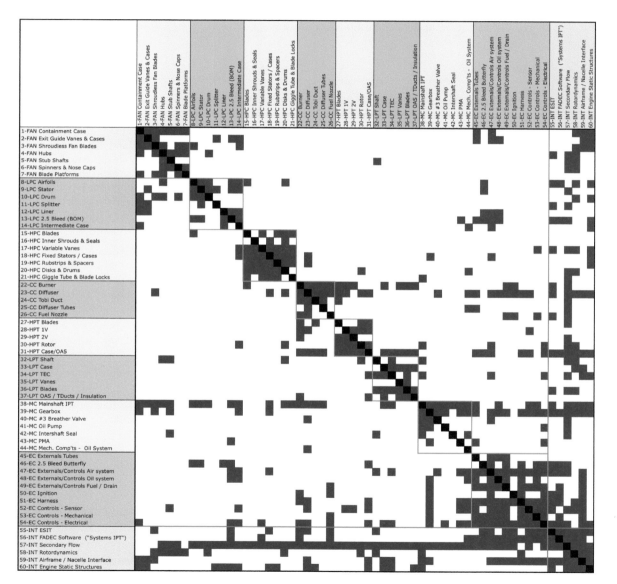

Figure 4.1
An organization architecture DSM model showing the network of communications across teams in a project at Pratt & Whitney developing a jet engine (see example 5.3).

In this chapter, we consider the architecture of organizations, with particular attention to organizations that develop engineered systems. We show how DSM is applied to represent and analyze such organizations and the types of insights gained through these DSM applications. We begin with a brief synopsis of terminology used in the particular context of organization architecture DSM modeling.

Terminology

Organization   A network of people with a common purpose, such as a business unit or a project developing, producing, selling, or supporting a product.

Organization Architecture   The structure of an organization—embodied in its people, their relationships to each other and to the organization's environment, and the principles guiding its design and evolution. Organization architectures generally group people into teams, departments, or other types of organizational units. The terms **organization architecture** and **organization structure** are often used interchangeably, although the latter term is also used in the more limited sense of lines of authority (reporting relationships).

Organizational Units   The elements comprising an organization, such as individuals, teams, groups, departments, and so on.

Interactions   The relationships among units in the organization. We are especially interested in information flow interactions, which may be formal or informal peer-to-peer communications, including e-mail, face-to-face discussions, group meetings, presentations, file transfers, and so on. Other interactions of interest in some cases may be based on relationships of authority, responsibility, accountability, contractual obligations, and so on.

Organization Architecture DSM   A mapping of the network of interactions among the people or units within an organization; also known as **organization DSM**, **organization structure DSM**, **people-based DSM**, and **team-based DSM**.

Cluster   A larger organizational unit (such as a department, team, or group of teams) suggested through analysis of the organization architecture DSM.

Integrative Mechanisms   The means by which work coordination and communication are facilitated across organizational units; also called **coordination mechanisms**.

Background

The effective development of products and systems requires project and program managers to facilitate the flow of information between people and across organizational units. This presents a dilemma for managers. On the one hand, the appropriate information must flow to the right people at the right times. Thus, managers may wish to enable more and better communication, the free flow of ideas, and the open sharing of issues and concerns, with hopes of building consensus and preempting problems. On the other hand, this can go too far: Sending everything to everyone in an organization can be problematic, leading to the familiar phenomenon of information overload. Many individuals in contemporary organizations receive hundreds of e-mails per day, and it is simply impossible to handle all of them. Many people send an e-mail and assume that it is read and understood. This is perhaps worse than not sending the information at all because, while successful communication has not occurred, the sender may believe it has. Meetings are another technique commonly employed to enable organizational communications. However, some people find that they spend most of their time in meetings, without much time left for actually doing work. So, an unregulated free flow of information is not the answer either. Managing the flow of information to facilitate the work of large organizations and complex projects is one of the key reasons that managers seek to design and architect organizations purposefully.

Organization architecture (or structure) has to do with the way people work together to deliver value—to accomplish the work of the organization. Specifically, organization architecture consists of three mappings depicted in figure 4.2: (1) hierarchical decomposition of the organization into elements (units) such as departments, teams, and individuals; (2) work assignments and top-down reporting relationships (lines of authority) within the organization; and (3) lateral relationships (especially information flow) among the organizational units, also called the interaction network. The first two mappings (decomposition and reporting roles) are often represented by an organization breakdown structure (OBS) diagram, commonly known as an *organization chart*. The DSM has been applied to the third of these mappings, which calls for a square matrix of interactions. Although many organizations consider their structure primarily in terms of the decomposition and defined roles, DSM models offer tremendous value through additional insights not provided by organization charts.

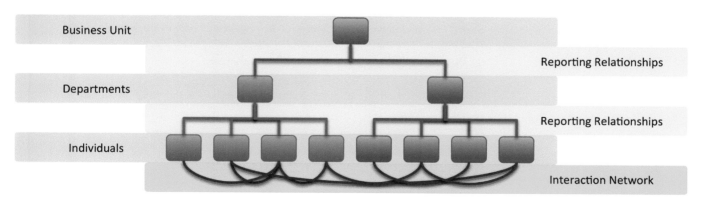

Figure 4.2
Organizations are typically decomposed into departments and other groups of people assigned to various roles such as projects. The network of interactions between people working on a project may be captured in an organization architecture DSM.

The understanding that organizations can be designed in superior ways, and that superior organization structures can provide competitive advantages, has motivated research in several areas. Organization science teaches us that organizations can be designed "rationally" based on a detailed understanding of the necessary flows of communication. These communications are what we represent in a DSM as the basis for analysis and/or design of the organization. Benefits of rational organization design include improved team structures and insight on the application of *integrative mechanisms* (or *coordination mechanisms*) (see, e.g., Galbraith 1994). Figure 4.3 lists several integrative mechanisms that have proven useful in the context of engineering projects.

The type of DSM used for organization analysis and design is called the *organization architecture DSM*, also known as the *organization DSM*, *organization structure DSM*,

1.	*Co-location*	Physical adjacency of organizational units (e.g., individuals and teams)
2.	*Traditional meetings*	Face-to-face gatherings for information sharing and/or decision making
3.	*Improved information and communication technologies*	Collaborative tools, e-mail distribution lists, tele- and videoconferencing, linked software systems for product design, shared databases, etc.
4.	*Training*	Team building (at each level of integration in the hierarchy), increasing awareness about integration needs and roles
5.	*Town meetings*	Not to share technical information, but to boost camaraderie, increase awareness of wider issues, and bolster the shared culture
6.	*Management mediators*	Orchestrators, integrators, and heavyweight managers (Clark and Wheelwright 1993)
7.	*Participant mediators*	Boundary spanners, liaisons, and conflict resolvers
8.	*Interface control groups*	Integration teams tasked with ensuring ongoing or incident-specific mediation of issues regarding specific interfaces
9.	*Standard process models*	Shared routines and procedures, specification of interfaces and metrics for evaluating interface effectiveness, interface contracts and scorecards (Browning et al. 2006)
10.	*Boundary objects*	Artifacts manipulated by those on both sides of an interface (Star and Griesemer 1989), such as shared models
11.	*Incentive systems*	Shared rewards and/or penalties for performance in relation to interfaces or other teams
12.	*Shared interpretations*	Common interpretations of design goals, objectives, and problems (Bernstein 2001), often from common backgrounds or experiences
13.	*Shared knowledge*	Common understanding and skill sets (Hoopes and Postrel 1999)
14.	*Shared ontologies*	Common terminology across teams for products, processes, and tools
15.	*Situation visibility*	Shared visual orientation of a team's activities and results in relation to other teams' activities in "the big picture" (Steward 2000)

Figure 4.3
Some integrative (or coordination) mechanisms (adapted from Browning 2009).

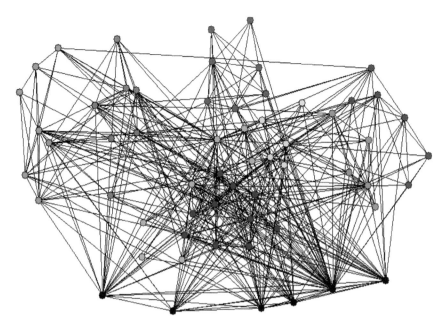

Figure 4.4
A communication network represented by a node-link diagram (undirected graph) instead of a DSM.

people-based DSM, or *team-based DSM*. This DSM captures the structure of organizational units and their interactions. The DSM represents people, teams, departments, or other organizational units as the diagonal cells (also naming the rows and columns) of the matrix. The communication pathways among these elements are captured by the marks or values in the off-diagonal cells.

Communication networks can also be depicted using a graph in which interactions among nodes (usually people) are drawn as a network of arcs. For example, figure 4.4 shows an undirected graph of the same communication network data represented in the DSM of figure 4.1. These types of node-link diagrams are common in systems analysis, including that of social networks. Although diagrams such as these have some benefits, a DSM view generally improves the layout of such information.

Square matrices had already been employed to represent organizational communication (Allen and George 1993; Lorsch and Lawrence 1972, p. 107) and other types of social networks before the term DSM became attached to them. However, to our knowledge, such matrices were not used as the basis for any specialized analyses aside from merely summing rows and columns. In 1993, McCord and Eppinger at MIT used a DSM to represent and analyze organization architecture (example 5.1). This research captured the frequency of interactions between teams in a large product development project at General Motors (GM) and prescribed an organization structure with improved modularity and means for integration and coordination.

Many problems in projects can be attributed to integration challenges across organizational units. For example, two teams may realize too late that they did not properly coordinate certain product design features, resulting in a problem that delays the project. Moreover, poorly coordinated teams can suffer from gaps (each thinking the other is doing something) or overlaps (redundant work). Therefore, a systematic approach is needed for considering coordination and integration up front when the organization is designed.

Structures within organizations such as teams and departments tend to facilitate internal communications but also may hinder external communications. Therefore, it is helpful to map and understand the information flows in the organization and to adjust the structure to facilitate the proper flows. For example, the assignment of a person to a team often has implications for physical location (e.g., co-location with other employees), e-mail distribution lists, and attendance at particular meetings. In other words, integrative mechanisms like these tend to be applied to the formal structure of groups designed into the organization. However, in the design of this formal structure, it is wise to consider the network of desired information flows so that the integrative mechanisms will be most effective. One way to achieve this in established organizations is to capture in the DSM some of the informal flows that have evolved among the workforce as channels for both routine work and problem solving. For newer organizations without such established flows, the DSM would be used instead to identify the network of desired information flow relationships based on individuals' expectations and experiences with the type of challenge at hand.

Building an Organization Architecture DSM

The basic procedure for building an organization architecture DSM model (which is similar to that used to create a product architecture DSM model) is as follows:

1. Decompose the overall organization into its elemental-level units, such as departments, teams, and/or individuals. (It is not uncommon for the resulting OBS to have many similarities to the corresponding product breakdown structure [PBS] because organizational units are often designated to be responsible for particular aspects of a project's desired result.) Lay out the square DSM with names of the organizational units labeling the rows and columns, grouped into higher-level organizational units if appropriate.

2. Identify the discovered (or desired) communication interactions between the units and represent these using marks or values in the DSM cells.

The GM engine development program DSM shown in figure 4.5 illustrates this basic procedure. The next chapter includes further explanation of this DSM application (example 5.1).

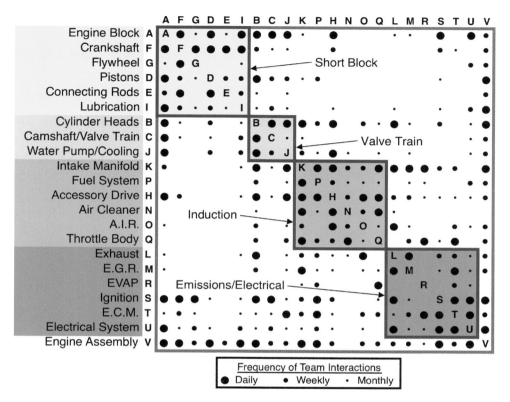

Figure 4.5
An organization architecture DSM model, representing engine development team interactions at General Motors (McCord and Eppinger 1993).

Figure 4.5 shows the organization architecture DSM representing the teams that developed a small-block V8 engine at GM in the 1990s. The DSM shows decomposition of the organization into 22 cross-functional component development teams (component teams) and the frequency (daily, weekly, or monthly) of communications reported between the teams. Each component team had responsibility for developing its component or subassembly of the engine, along with its associated production system. The 22 component teams were initially grouped into four subsystem engineering teams charged with integration of the engine components and delivery of overall, system-level performance.

The type of organization structure depicted in figure 4.5 is actually quite common in the development of complex engineered systems (as the examples in chapter 5 illustrate). Leaving aside the actual GM situation for a moment, let us consider the implications of this type of organization structure.

1. Suppose that each component team meets weekly, whereas each subsystem team holds a bimonthly meeting attended by at least each component team's leader. This is one type of integrative mechanism that might be used to facilitate the proper dissemination of information and, hopefully, to enable any cross-team issues to surface sooner rather than later. However, the intensity of the interactions outside the formal organizational structure (outside the shaded squares in the DSM representing the subsystem team boundaries) suggests that this approach is unlikely to prompt all of the desired integration across teams.

2. Suppose that each component team appoints a liaison to the other teams in its subsystem team. Although this might be helpful, it would not address integration issues with the component teams in other subsystem teams.

3. Suppose that the project leaders realize that not everyone on every team can be copied on all e-mails, so they set up distribution lists based on the subsystem team assignments. Again, this arrangement seems unlikely to ensure sufficient coordination among all of the teams.

4. Suppose that the project is actually able to physically co-locate all of its teams in an office building. Managers may choose to assign office areas based on the subsystem team groupings. While facilitating the communications within each subsystem team, this decision would inhibit communications across subsystem teams, increasing the likelihood of integration issues surfacing later at these fissures.

5. Suppose that the program decided to work with some major component suppliers. Which components could be readily outsourced for development by the suppliers? Indeed, every component team has significant communication needs with several other teams. Therefore, any outsourced component development would likely create even more challenging and problematic coordination within the project.

Hence, a project's organizational architecture has major implications for its ability to apply—and the effectiveness of—various integrative mechanisms. Where communication needs exist, they can be either facilitated or inhibited by the application of such mechanisms.

Here are several caveats to consider when building organization architecture DSM models:

• **Granularity**   The level of decomposition into organizational units determines the granularity or richness of the DSM model. Organization DSM analysis is most often done at the level of teams or individuals. However, for large organizations or projects, the analysis may be done with departments or business units. Of course, this is also a tradeoff with the usability of the model. We have generally found that models on the order of 20 to 50 units are highly understandable and most useful.

- **Data collection**   To document communication interactions for the DSM, it is helpful to focus data collection with perhaps a single question. Most commonly, this has been either frequency of communications between the organizational units and/or importance of such interactions. When collecting data from a team, it is often expedient to query the team leader, although this individual may lack full knowledge of the team's interactions with other teams. In that case, it can be helpful to get a second opinion or ask the full team to verify the team leader's responses.

- **Symmetry**   Organization DSM models are often symmetric because if one person (or team) communicates with another, the interaction is often reciprocal. However, it is not uncommon for a majority of the actual information flow to occur in just one of the directions between two parties. If the modeler wants to distinguish directionality of flow, then it is important to capture it as part of the data collected.

- **Accuracy**   Because they tend to be symmetric, each interaction in the DSM should typically be noted by two respondents representing both parties to the communication. However, when building the DSM, it is common to find several one-sided or mismatched interactions (i.e., instances where one party reports the communication and the other does not, or one party rates the interaction as more frequent or more critical than does its counterpart). When the responses differ, this can usually be resolved by bringing both sides together to discuss the nature of their interactions. We have found that in most cases when an interaction is overlooked or underrated by one of the parties, it does in fact exist, although one of the respondents was not fully aware of it; only rarely does a respondent insist that a disputed interaction does not in fact occur. We have found that merely building and verifying the DSM model provides the valuable benefit of reconciling the various respondents' flawed mental models of the organization's information flows.

- **Representing interactions**   The strength of the interactions may be represented in the DSM by numerical values, letters, colors, or graphical elements (or some combination thereof). Although numbers can be useful for further analysis, visualization is best accomplished with shading or graphical symbols (such as those used in figure 4.5).

- **Dynamics**   Organization DSM models generally provide a static description of the information flow within an organization. However, project-based organizations are by nature highly dynamic. Individuals may be assigned to different projects or roles over time, group assignments may change, communication needs may change as the work progresses, and so on. Therefore, it is wise to consider the time frame desired in the model when collecting the data. Separate organization DSMs may be built at periodic intervals and compared to increase understanding of project and organizational dynamics (see examples 5.2 and 9.12).

Analyzing the Organization DSM

The analytical methods applied to organization architecture DSM models are quite similar to those described for product architecture DSMs in chapter 2, so our description here focuses on the minor differences in techniques and interpretation. Because organization DSMs are static (i.e., all of the organizational units represented in the matrix exist simultaneously), clustering analysis (both manual and automated in software) is typically used to assign organizational units to groups.

Engineers find the analogy between product architecture and organization structure to be quite useful. If the units of our organization architecture DSM analysis are the individuals in the organization, then these are analogous to components in the product architecture DSM. The clusters or groups of people in the organization are then akin to the product modules. We can now apply the same clustering techniques used for product architecture DSMs; the primary difference may be the formulation of the objective function.

Typically, the objective of clustering an organization DSM is to assign the people having the greatest needs to communicate with each other to the same groups because this designation often implies the natural application of integrative mechanisms such as co-location, meetings, distribution lists, and managerial oversight, as discussed earlier. It is best to avoid reliance on critical communication across groups because such paths may lack natural communication facilitators. However, putting everyone in a large organization into a single group is also undesirable because this amounts to telling everyone to communicate with everyone else without any specific guidance, which would undoubtedly lead to information overload. Any critical communications that must take place across groups become targets for special managerial attention and further integrative mechanisms.

Figure 4.6 shows the results of the manual clustering analysis used to reorganize the GM engine project into five more natural groups based on the communication data in the DSM. Here, component team assignments to groups were based on the reported frequency of their interactions. The premise underlying this clustering approach is that teams needing to exchange information frequently would benefit from tighter integration through the formal organization structure of higher-level groupings. Where necessary, component teams were assigned to two or three subsystem teams. Thus, by the nature of their interactions, some component teams needed to be part of more than one e-mail distribution list, attend more system integration meetings, appoint more liaisons, and so on. The five component teams grouped at the bottom of the matrix did not fit neatly into the four subsystem teams. These five component teams needed to interface with practically all of the others. Therefore, these five teams were grouped into an integration team. To implement this structure, GM managers asked a representative from each of these teams to attend each of the other four groups' meetings. Whatever integrative mechanisms were deemed most appropriate, their application would be more effective when based on the underlying needs for coordination dictated by the organizational architecture.

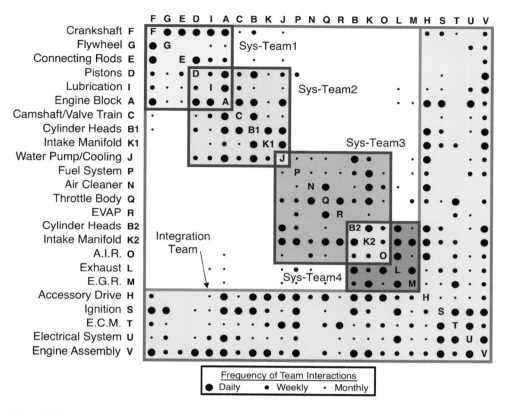

Figure 4.6
Modified GM engine development organization structure suggested by the DSM clustering analysis.

Usually the analysis of an organization DSM explores several scenarios for the organization design, trading off the pros and cons of assigning various units to particular clusters. The analyst may need to consider physical or political constraints on the size and composition of groups, such as the size and/or location of a facility or established reporting relationships. The menu of available integrative mechanisms plays a large role in enabling a broader range of effective scenarios, such as communication technologies that compensate for a lack of co-location, dedicated liaison roles to coordinate specific interactions, and so on. The presence or absence of such options will render certain organizational architectures more or less desirable.

Other questions to consider during the analysis include the following:

- Should some units be aggregated or divided? Perhaps the initial OBS needs revision. Perhaps two teams within a cluster could be combined. Or, perhaps a team that could fit well in either of two clusters could be divided or left intact but assigned to both clusters.

- What about communication that occurs across organizational levels?

- Should a group of teams designate another team as an input/output manager? Sometimes a single team can serve as the liaison or conduit for a cluster of other teams' external interactions.

- What about a team whose job is to broadcast communication? This tends to be an integrative role that will probably fit best in an "integration team" like the one in the GM example.

Applying the Organization Architecture DSM

Organization architecture DSM models have been applied to both analyze existing organization structures and their communications and plan new organization structures. The models have produced many useful insights leading to redesign of organizations or identification of areas where integrative mechanisms should be applied. Many organization architecture DSM examples are given in the next chapter. Typical applications include:

- **Application of integrative mechanisms**   Some types of integrative mechanisms are easily scaled to large organizations (e.g., database access, e-mail distribution). Other mechanisms are best applied to small groups (e.g., face-to-face team meetings, liaisons). Organization DSM analysis can assist in choosing appropriate mechanisms for various coordination challenges within a large project (see examples 5.1, 5.2, 5.3, 5.4, 5.6).

- **System engineering teams**   The most common usage to date of organization DSM analysis has been to assess the communication needs across team members in a large engineering project. The boundaries of subsystem and integration teams can be planned based on the interactions in the DSM (see examples 5.1 and 5.3).

- **Definition of project teams**   To plan a single project team within a larger organization, the DSM is able to identify the core team members based on the communication needs of the project work (see example 5.6).

- **Rational organization design**   The premise of this approach is that the people in the organization need to communicate because the work they are doing is somehow related. This type of rational organization design may be based on either the product architecture or the process architecture, represented in DSM format, by assigning organizational responsibilities to each product or process element. This is a way to design entirely new organizations (because it would be difficult to assess the necessary communication patterns without people to ask) (see examples 5.1, 5.6, 9.3).

- **Facility layout**   By identifying the communication needs across people and departments, it is possible to determine an efficient assignment of units to office locations (see example 5.5).

Conclusion

The organization architecture DSM has proven to be an effective representation for a system of organizational units and their relationships. It provides an intuitive visualization tool, facilitating discussions and insights about information flow patterns and their implications. Typically, just building a DSM forces disparate people, teams, and groups to increase mutual awareness and understanding. Merely aligning people's mental models of the information flow relationships and patterns can add tremendous value in organizations. Furthermore, organization DSM models can be analyzed via clustering, which remains somewhat of an art and is generally combined with manual manipulation of the DSM to generate alternative perspectives on the organization architecture. These models improve organizational understanding, facilitate organizational innovation, and inform the appropriate application of integrative mechanisms.

The value of this type of DSM increases as organizations become larger and more complex. Coordination can be managed informally in small organizations but breaks down in large ones. Indeed, two of the main benefits of a DSM model are its abilities to (1) concisely represent a relatively large number of organizational units and their relationships, and (2) identify important groups of organizational units and patterns of interactions. The DSM helps maintain a shared understanding of the organizational architecture and its implications.

References

This list of references provides additional background on the organization architecture DSM.

To our knowledge, the 1972 book by Lorsch and Lawrence (p. 107) contains the earliest published example of a square matrix mapping the relationships between organizational units.

Lorsch, Jay W., and Paul R. Lawrence, eds. 1972. *Managing Group and Intergroup Relations*. Homewood, IL: Richard D. Irwin.

Allen studied organizational communications using several methods, including graphs depicting the networks of connections. Later with George, he used square matrices called netgraphs to represent sampled data of daily communications between individuals.

Allen, Thomas J. 1977. *Managing the Flow of Technology*. Cambridge, MA: MIT Press.

George, Varghese, and Thomas J. Allen. 1993. Relational Data in Organizational Settings: An Introductory Note for Using AGNI and Netgraphs to Analyze Nodes, Relationships, Partitions and Boundaries. *Connections* XVI (1 & 2).

McCord and Eppinger developed the first application of DSM to a network of team interactions (example 5.1). (Prior DSM work had been limited to process- and parameter-based models.)

McCord, Kent R., and Steven D. Eppinger. 1993. *Managing the Integration Problem in Concurrent Engineering.* MIT Sloan School of Management, Working Paper no. 3594.

Eppinger, Steven D. 1997, August. *A Planning Method for Integration of Large-Scale Engineering Systems.* International Conference on Engineering Design, Tampere, Finland, pp. 199–204.

Extending his 1996 thesis and a 1999 paper, Browning discussed issues pertaining to multiteam integration in large projects or programs. He described a process called "design for integration," whereby managers could architect a project organization in light of the product architecture and with the application of 15 types of integrative mechanisms.

Browning, Tyson R. 2009. Using the Design Structure Matrix to Design Program Organizations. In *Handbook of Systems Engineering and Management*, 2nd ed., eds. Andrew P. Sage and William B. Rouse. New York: Wiley, pp. 1401–1424.

Two articles by Sosa, Eppinger, and Rowles derive from a unique study comparing the network of component interfaces (product architecture DSM) to the network of team interactions (organization architecture DSM) for a jet engine development project. The *Management Science* article explored several explanations to help understand the misalignment between the two architectures. The *Journal of Mechanical Design* article defined the use of the terms *modular* and *integral* to describe subsystems within complex system architectures and identified the impact of these structures on the design team interactions.

Sosa, Manuel E., Steven D. Eppinger, and Craig M. Rowles. 2003. Identifying Modular and Integrative Systems and Their Impact on Design Team Interactions. *Journal of Mechanical Design* 125 (2):240–252.

Sosa, Manuel E., Steven D. Eppinger, and Craig M. Rowles. 2004. The Misalignment of Product Architecture and Organizational Structure in Complex Product Development. *Management Science* 50 (12):1674–1689.

These additional references pertain to the integrative mechanisms listed in figure 4.3.

Bernstein, Joshua I. 2001. *Multidisciplinary Design Problem Solving on Product Development Teams.* PhD thesis (TMP), Massachusetts Institute of Technology, Cambridge, MA.

Browning, Tyson R., Ernst Fricke, and Herbert Negele. 2006. Key Concepts in Modeling Product Development Processes. *Systems Engineering* 9 (2):104–128.

Clark, Kim B., and Steven C. Wheelwright. 1993. *Managing New Product and Process Development.* New York: Free Press.

Galbraith, Jay R. 1994. *Competing with Flexible Lateral Organizations.* 2nd ed. Reading, MA: Addison-Wesley.

Hoopes, David G., and Steven Postrel. 1999. Shared Knowledge, Glitches, and Product Development Performance. *Strategic Management Journal* 20:837–865.

Star, Susan L., and J. R. Griesemer. 1989. Institutional Ecology, "Translations." and Boundary Objects: Amateurs and Professionals in Berkeley's Museum of Vertebrate Zoology 1907–39. *Social Studies of Science* 19 (3):387–420.

Steward, Donald. 2000. *A Very Brief Discussion of Information Driven Business Management: A Problem Solving Approach,* <http://www.problematics.com/readings.asp>, Last accessed May 26, 2011.

5 Organization Architecture DSM Examples

Overview

This chapter presents seven example applications of the organization architecture DSM as listed in the table below. Each example describes the purpose of the model (problem to be addressed), how the data were collected, how the model was built, and the results. References for further information, where available, are also provided.

Example	Application	Organization	Purpose
5.1	Automobile engine development project	General Motors, USA	▪ Redesign organization architecture for enhanced communication and integration
5.2	Military aircraft development program	McDonnell Douglas, USA	▪ Understand the program's organizational architecture and dynamics
5.3	Commercial aircraft jet engine development project	Pratt & Whitney, USA	▪ Investigate patterns of organizational communication within and across subsystem teams
5.4	International Space Station	NASA, USA	▪ Evaluate sustaining engineering strategy, critical skills, communication, and coordination
5.5	R&D Center	Timken, USA	▪ Plan arrangement of offices in new technology center
5.6	LNG terminal development project	BP, UK	▪ Improved organization of a large engineering project
5.7	Multinational energy project stakeholders	BP, UK	▪ Analyze the stakeholder value network ▪ Identify channels of stakeholder influence

Example 5.1   General Motors Powertrain V8 Engine Development

Contributors

Steven Eppinger and Kent McCord
Massachusetts Institute of Technology

Problem Statement

Development of a small-block V8 engine at General Motors Powertrain Division in 1992 was organized as a network of teams. This organization architecture application of DSM was aimed at improving the effectiveness of GM Powertrain's system engineering process by enabling more direct and explicit communication within and across the teams. The engine development project consisted of 22 cross-functional component development teams (component teams [CTs]) grouped into four subsystem engineering teams (STs).

Data Collection

We started with decomposition of the engine (shown in figure 5.1.1) into its 22 major components or subassemblies. The organization structure corresponded directly to the product decomposition, resulting in 22 CTs, each responsible for design and development

Figure 5.1.1
General Motors small-block V8 engine (courtesy of General Motors).

of one major component or subassembly and its production system. We then asked the leader of each CT to complete a simple one-page survey form, indicating how often their CT needed to work with each of the other CTs (daily, weekly, or monthly) in the detailed design phase of the project. We then identified which CTs comprised each of the four STs and arranged the DSM accordingly before conducting our own clustering analysis.

Model

The raw data DSM in figure 5.1.2 shows how frequently each of the 22 CTs reportedly worked with the others. The clusters shown in the DSM of figure 5.1.3 indicate the original assignment of the CTs to STs. Our clustering analysis sought an alternative organization architecture that would group the CTs into more effective STs such that more of the CT

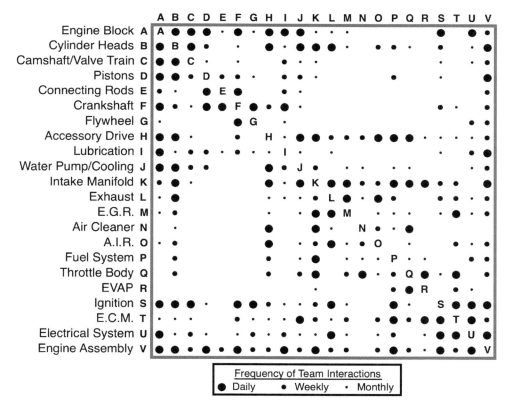

Figure 5.1.2
Team communication data set.

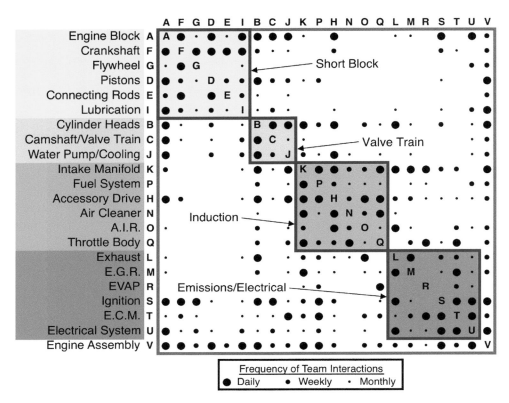

Figure 5.1.3
Original system team structure.

interactions would be within STs and fewer would take place outside of the system team structure. Figure 5.1.4 shows the results of our reclustering and the proposed reorganization of the project.

Results

This was perhaps the first time that a complex technical project at GM was organized based on data representing their own communication needs. The original ST organization comprised four STs (Short Block, Valve Train, Induction, and Emissions and Electrical). Each ST would meet every two weeks to discuss the integration of their components to deliver system-level performance. However, the initial DSM layout indicates that this structure enabled only some of the dozens of interactions that needed to take place across CTs. We asked the program managers how they address the interactions that are not within the STs, and they told us that many of the interactions may not in fact be addressed

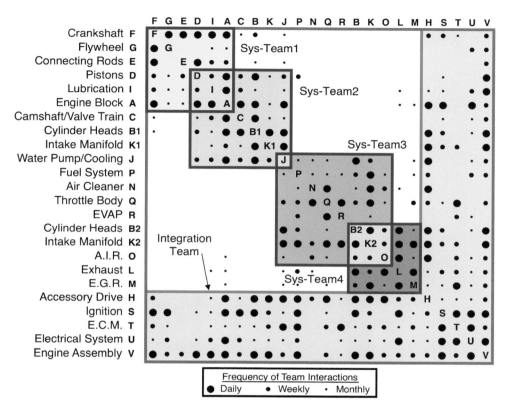

Figure 5.1.4
Proposed system team structure.

until a problem arises, potentially much later in the system integration phase of the project. They told us they would appreciate our help to improve this situation.

Our DSM clustering analysis suggested four STs and one integration team (IT). Each ST comprised several of the CTs (as before); however, now each CT was assigned to one *or more* of the STs or to the IT. This structure greatly reduced the number of interactions occurring outside of the ST structure.

Note that in the proposed reorganization, some CTs were assigned to two STs (Pistons, for example). Two CTs (Cylinder Heads and Intake Manifold) were each assigned to three of the STs. These assignments, of course, reflected each CT's need to interact with certain of the other CTs and vice versa. Finally, five of the CTs essentially reported that they needed to work with almost all of the other CTs, and so these five were assigned to be the IT.

Implementation of the proposed system engineering organization structure at GM Powertrain was fairly straightforward. First, they adopted the new structure of the STs and planned the ST meetings to be on different days of the week to accommodate those CTs on multiple STs. Second, the IT was given the responsibility to (1) check in with each of the STs on a regular basis (and to attend their meetings as needed), (2) meet as an integration team to address system-level engine integration and performance issues, and (3) help the program managers to direct the final system integration phase of the development process.

General Motors Powertrain reported to us that the small-block V8 engine development program that was the subject of this example had the "smoothest integration phase ever." They attributed this result to a great extent to the new system engineering team structure shown here.

References

Eppinger, Steven D. 1997, August. *A Planning Method for Integration of Large-Scale Engineering Systems*. International Conference on Engineering Design, Tampere, Finland, pp. 199–204.

Eppinger, Steven D. 2001, January. Innovation at the Speed of Information. *Harvard Business Review* 79 (1):149–158.

McCord, Kent R. 1993, August. *Managing the Integration Problem in Concurrent Engineering*. MIT Sloan School of Management, Working Paper no. 3594.

Example 5.2   McDonnell Douglas F/A-18E/F Program

Contributor

Tyson Browning
Neeley School of Business, Texas Christian University

Problem Statement

The Boeing F/A-18E/F Super Hornet is a fighter/attack aircraft originally developed by McDonnell Douglas for the U.S. Navy (figure 5.2.1). The E/F program constituted a major redesign of the earlier (A-D) versions of the aircraft. A 1995 study investigated the integrative mechanisms among the program's cross-functional development teams during the Engineering Manufacturing Development (EMD) phase (which spanned late 1992 to 1996). Part of that study entailed building two "quick-look" DSM models of part of the program's organizational architecture. One of the models focused on the current organization, whereas the other examined an earlier situation for comparison.

Data Collection

The leader of each of 41 teams received a two-part survey, one to respond about the current situation and the other to respond retrospectively about the situation 18 months earlier. Twenty-three of 41 team leaders responded (56% response rate); the program's limited resources precluded additional follow-up. However, the "quick-look" purpose of the model was satisfied. Each team leader was asked to indicate whether they provided and/or received program information from each of the other 40 teams. If yes in either case, they were further asked to rate the frequency of the interaction on the following scale: 1 = infrequent (monthly), 2 = frequent (weekly or biweekly), and 3 = regular (daily).

Model

The DSM shown in figure 5.2.2 represents the initial (1995) situation for the 23 teams that responded. The teams were grouped according to the program's OBS. The off-diagonal cells show the reported frequency of technical information transmission from the team in row i to the team in column j. (Note that this is the input-in-columns [IC] convention, the transpose of the matrix convention used in the other examples in this chapter.) This DSM is actually the composite of two DSMs, one representing the information provider point of view (i.e., built row by row) and the other the receiver perspective (i.e., built column by column). These two responses should have been identical but were not

Figure 5.2.1
F/A-18E/F Super Hornet (courtesy of Boeing).

always because sometimes one team leader would indicate that his team provided information to another team with one frequency, whereas the leader of the other team indicated that his team received it with another frequency. Where the responses agreed, the DSM in figure 5.2.2 shows the off-diagonal cell in white (no shading). Interestingly, a common perception of the interaction was the exception rather than the norm. More often, the provider and the receiver's responses did not agree. Where the two responses differed by only one level, the DSM shows their average, so they appear with a 0.5 appendage in a yellow-shaded cell. For example, if one team leader said the output was daily (3) while the other said the input was weekly (2), then the DSM shows 2.5. A number of the responses differed more substantially. The 2s and 3s in the red-shaded cells represent instances where one team leader said 0 or 1 and the other said 2 or 3 (i.e., responses differing by two or three levels, respectively). Ideally, these discrepancies would invite follow-up to determine the source of the misunderstanding. Although follow-up was not feasible in this case, results from similar models indicate that, once a discrepancy is highlighted and discussed by the affected teams, in most cases they find that (1) there really is an interaction occurring, and (2) the actual frequency is the greater of the two reported frequencies. Thus, a corrected, final DSM model can be approximated by taking the maximum of the two responses in each cell.

The resulting DSM is shown in figure 5.2.3, where the size of the dot replaces the numbers (to aid in visualization). Extra columns to the right of and below this DSM tally

	A	B	C	D	E	F	G	H	I	J	K	L	M	N	O	P	Q	R	S	T	U	Total	Avg.
Inner/Outer Wing **A**	A	2.5	1.5		.5	2.5	1.5				.5	.5	.5		1			2		3	3	19	1.0
LE Flaps/Horizontal Tail **B**	2.5	B	3		1	2.5	2					.5	.5			1.5				3	3	19.5	1.0
TE Flap/Aileron **C**	1.5	3	C		1	2	2	.5	.5		1	1	1		.5					3	3	20	1.0
Maneuvering Loads **D**	2	1.5	1.5	D	2	2.5	2	2	1.5		1	1.5	1.5	1	1	1.5	1	2	.5	1	.5	27.5	1.4
E & B Loads **E**	1	1	1	2	E	2.5	1.5				.5	1	.5	.5	.5	.5	1.5	2	2			18	0.9
Structural Dev. & Test **F**	2.5	2.5	2	3	2.5	F	2.5					.5		2	1.5	1		2		2	2	26	1.3
Structural Integrity **G**	1.5	2	2		2	2.5	G		.5			.5		2	1.5	.5		1.5		1.5		18	0.9
Flt Ctrls Computer Soft. **H**				2				H	3	3		.5						.5	2			11	.6
Flying Qual/Control Laws **I**			.5	1.5	1		.5	3	I	2.5	.5	2		.5		1						13	.7
Flt Ctrls Syst Integ Testing **J**			.5					3	2.5	J						.5		1	2			9.5	.5
Weapons Separation **K**	.5		.5	1	.5		.5		1		K	1.5	2					2.5				10	.5
Stability & Control **L**	.5	.5	1	1.5	1			2			1.5	L	2.5					3		.5		14	.7
High Speed Drag & Perf. **M**	.5	.5	1	1.5	1	.5	.5	.5	1.5		2	2.5	M			1	2	2		.5		17.5	.9
Main Landing Gear **N**						2	2		.5		.5			N	2.5	1.5		2	.5	3		14.5	.7
NLG/Doors/Hooks **O**						2	2							2.5	O	1	1	2			.5	11	.6
Mechanisms/Flt Controls **P**	1	1.5	.5		.5	1	.5	1	.5				1	1.5	1	P		.5	2	2.5		15	.8
ECS **Q**	.5	.5	.5		2		.5						2		1	.5	Q		2.5	.5	1	11.5	.6
Armament **R**	2	.5	1	2	2	2	1.5	.5		1	2.5	3	2	2	2	.5		R	2	3		29.5	1.5
Electrical **S**								2		1.5				.5	.5	2	2.5	2	S	2		15	.8
Assembly AT **T**	3	3	3			2	2							3		2.5		3	2	T	3	26.5	1.3
Composite Center AT **U**	3	3	3			2									.5		.5	3			U	15	.8
Total:	22	22	23	15	19	26	22	11	14	9	10	12	17	16	12	17	9	28	16	29	16		
Avg.:	1.1	1.1	1.1	0.7	1.0	1.3	1.1	.6	.7	.5	.5	.6	.9	.8	.6	.9	.4	1.4	.8	1.4	.8		

Figure 5.2.2
DSM showing initial data from team leader responses regarding frequencies of interactions with other teams in 1995.

the sum and average for each row and column. This DSM is also shaded to show the hierarchy of the organization (the blocks along the diagonal) and areas of especially intense interaction outside the current organizational structure.

The DSM shown in figure 5.2.4 represents the results of the second part of the survey (the situation 18 months prior, as recollected by the respondents, based on the teams existing at that time) after similar adjustments. Because each organization DSM shows a snapshot in time, a series of DSMs is needed to model discrete steps in organizational evolution. This comparison with an earlier stage of the program shows how teams can be added (e.g., teams T and U in the prior DSMs) and subtracted (e.g., teams A-E in the below DSM reduced to teams A-C in the above) and how relationships between teams can change (e.g., reduction in the frequency of interactions of teams A-C with teams Maneuvering Loads and Structural Integrity). Note also the reduction in overall intensity

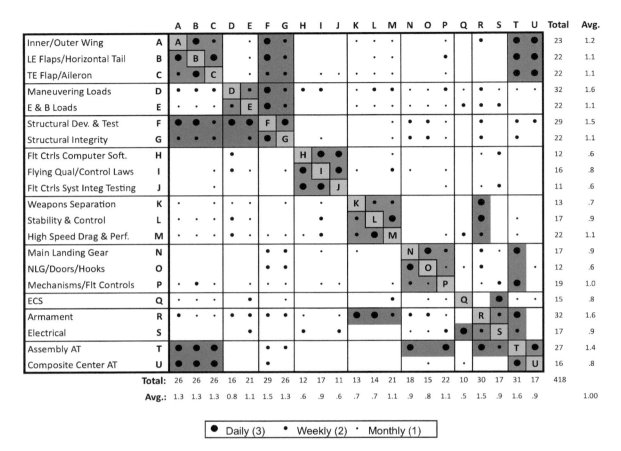

Figure 5.2.3
DSM showing organizational hierarchy and inferred frequencies of interactions in 1995.

of interaction (density of the DSM) as indicated by the drop in overall average interaction from 1.28 to 1.00.

Results

Even as incomplete models of the program's organizational architecture (because several teams are missing from the picture and several of the responses did not agree), these "quick-look" DSMs nevertheless provided a basis for several insights (without any clustering analysis). Two of these are described here.

First, it was clear that, at least initially, the team leaders did not really know who their team members interacted with and how often. Although a difference of one level, even

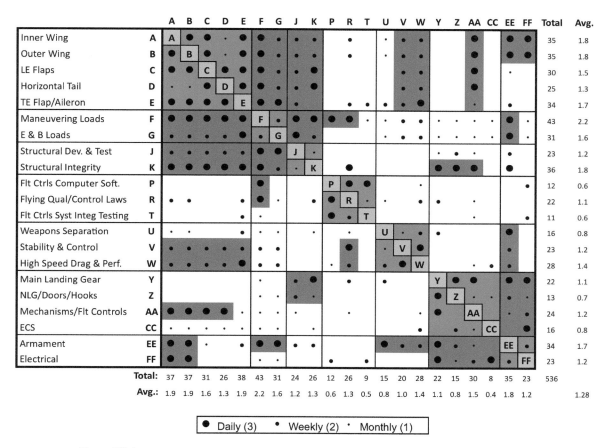

Figure 5.2.4
DSM showing organizational hierarchy and inferred frequencies of interactions in 1993 (from retrospective responses).

between 0 and 1, might not cause much concern, the larger discrepancies (e.g., between 0 and 3) are more problematic. The results would most likely have improved consistency if the team leaders had consulted their entire team before finalizing their responses, and follow-up could have addressed any misunderstandings. However, building the DSM exposed the team leaders' overall lack of interteam awareness.

Second, the frequent interactions among teams in the same part of the organization structure (i.e., interactions within the lightly shaded blocks along the diagonal in the DSMs) provide some justification for this organization design. However, the large number of interactions outside these main-diagonal blocks (shaded to highlight groups of especially intense connections) imply the need for additional integrative mechanisms besides the meetings, e-mail distribution lists, and physical co-locations that tend to mirror the organizational structure. Other integrative mechanisms such as collaborative tools, targeted meetings, appointed liaisons, joint team members, standard processes, boundary objects, or shared ontologies might be helpful in such cases.

References

Browning, Tyson R. 1996. *Systematic IPT Integration in Lean Development Programs.* Master's thesis (Aero.-Astro./TPP), Massachusetts Institute of Technology, Cambridge, MA.

Browning, Tyson R. 1998. Integrative Mechanisms for Multiteam Integration: Findings from Five Case Studies. *Systems Engineering* 1 (2):95–112.

Browning, Tyson R. 1999. Designing System Development Projects for Organizational Integration. *Systems Engineering* 2 (4):217–225.

Browning, Tyson R. 2009. Using the Design Structure Matrix to Design Program Organizations. In *Handbook of Systems Engineering and Management*, eds. Andrew P. Sage and William B. Rouse. New York: Wiley, pp. 1401–1424.

Example 5.3   Pratt & Whitney Jet Engine Development

Contributors

Manuel Sosa
INSEAD

Steven Eppinger
Massachusetts Institute of Technology

Craig Rowles
Pratt & Whitney

Problem Statement

Pratt & Whitney, a division of United Technologies Corporation, produces and supports commercial and military aircraft jet engines, industrial gas turbines, and space propulsion systems. Development of a commercial aviation jet engine is a highly complex process involving hundreds of engineers working simultaneously on the various components and subsystems. This DSM application investigated the system engineering and system integration aspects of the engine development project through an organization architecture DSM model.

Data Collection

Over a period of four months in 1998, Craig Rowles (both an employee of Pratt & Whitney and a student in MIT's System Design and Management master's program) interviewed lead engineers of the teams responsible for the design of all major physical and functional engine components in the PW4098 engine program. Subsequent data codification, analysis, and interpretation of the DSM model were done jointly with Manuel Sosa, then a doctoral student at MIT working with Professor Steven Eppinger.

Model

The DSM model shown in figure 5.3.1 represents the formal and informal organization architecture of the design phase of the PW4098 commercial engine program. The organization was formally structured into 60 teams. Fifty-four component teams were responsible for the design of the 54 major engine components; these 54 teams label the first 54 elements of the DSM. These component teams are grouped into eight clusters, each corresponding to a subsystem (listed starting from the front of the engine and the top of the matrix): Fan, Low-Pressure Compressor, High-Pressure Compressor, Combustion

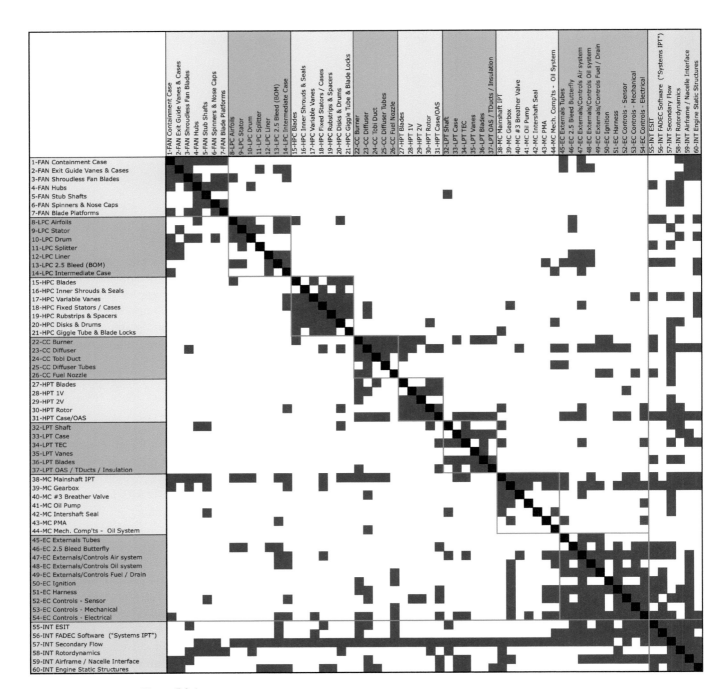

Figure 5.3.1
Organization architecture DSM for the PW4098 jet engine organization.

Chamber, High-Pressure Turbine, Low-Pressure Turbine, Mechanical Components, and Externals and Controls. The last six teams in the DSM correspond to six integration teams that had no direct hardware-design roles but were responsible for ensuring delivery of engine-level design requirements such as rotor dynamics and secondary air flow. A blue-colored cell (i, j) in the DSM indicates that team i acquired technical information (during the Design phase) from team j.

Results

The DSM model not only captured the formal organizational structure of design teams into eight subsystem groups but also captured the technical communication patterns of design teams within and across such groups. Identifying the technical communication patterns both within and across subsystem groups helped the engineering managers better manage the complex system engineering challenge. The system engineering practice had been largely focused on facilitating the communications inside the organizational boundaries.

Based on this analysis, they realized that a significant amount of important technical communication must occur across these boundaries. Managers decided to dedicate more attention to facilitate cross-team communication across organizational groups. For example, some of the cross-boundary team interactions between modular systems in our study were critical design interfaces that had not been previously identified by design experts. As a result of our study, managers learned about these interdependencies and established dedicated design teams or formally extended the responsibility of existing teams to explicitly handle these critical cross-boundary design interfaces during the development of the next engine.

For DSM analysis of the product architecture of the PW4098 engine, refer to example 3.2. A comparison of the product architecture DSM with the organization architecture DSM is presented in example 9.2.

References

Rowles, Craig M. 1999, February. *System Integration Analysis of a Large Commercial Aircraft Engine*. Master's thesis, Massachusetts Institute of Technology, Cambridge, MA.

Sosa, Manuel E. 2000, June. *Analyzing the Effects of Product Architecture on Technical Communication in Product Development Organizations*. PhD thesis, Massachusetts Institute of Technology, Cambridge, MA.

Sosa, Manuel E., Steven D. Eppinger, and Craig M. Rowles. 2004, December. The Misalignment of Product Architecture and Organizational Structure in Complex Product Development. *Management Science* 50 (12):1674–1689.

Sosa, Manuel E., Steven D. Eppinger, and Craig M. Rowles. 2007, November. Are Your Engineers Talking to One Another When They Should? *Harvard Business Review* 85 (11):133–142.

Example 5.4   NASA International Space Station Sustaining Engineering

Contributor

Tim Brady
NASA Johnson Space Center

Problem Statement

The International Space Station (ISS) began construction in 1998 and has had continuous occupation by human crews since November 2000 (figure 5.4.1). As the ISS grew, NASA began to plan for providing long-term engineering expertise to support sustained operations of the vehicle. In 2003, Kathy Lueders of the ISS Vehicle Office asked Tim Brady to evaluate the ISS sustaining engineering strategy.

Data Collection

Over the course of four months in 2003, Tim Brady (a NASA employee supporting ISS) reviewed ISS documentation and several independent studies of the planned long-term operation of ISS to characterize the technical effort required to provide the necessary

Figure 5.4.1
The International Space Station (ISS) in orbit (courtesy of NASA).

engineering expertise. In addition, interviews were conducted with more than 20 people in varied roles supporting on-orbit operations of the ISS and the Space Shuttle. The purpose of the data collection was to identify specific tasks and critical skills required for vehicle operations, examine organizational responsibilities, and examine information sharing, knowledge capture, and interaction among teams supporting ISS.

Model

Thirty-six critical functions performed by various teams within the ISS organization were identified to represent the scope of effort supporting on-orbit operations. These functions were placed into an organization DSM, and interdependency between functions was valued at 0 for no dependency, 1 for a function with moderate dependency on another, and 2 for high dependency. The organization DSM for these sustaining engineering functions is shown in figure 5.4.2. The color scheme (using conditional formatting in Excel) highlights the greater values.

A major theme captured from interviews related to the importance of critical skills retention. A second DSM (figure 5.4.3) was generated to analyze the critical skills used by the team members supporting ISS operations. This critical skills DSM was generated by first adding a row/column next to the functions list in the organization DSM. Each of the 36 functions in the DSM was assigned a weighting factor representing its critical skill value (CSV).

CSV	Criteria	Example Function
1	Requires general engineering or project skill	Perform configuration control
2	Requires skill unique to NASA	Test hardware (to NASA requirements)
3	Requires ISS-unique skills	Analyze ISS vehicle performance

The final step in forming the critical skills DSM was to calculate the value to be placed in each cell of the DSM based on the following formula:

$$\frac{\text{CSV of}}{\text{function A}} \times \frac{\text{CSV of}}{\text{function B}} \times \frac{\text{Functional dependence}}{\text{DSM value A-B}} = \frac{\text{CSV DSM}}{\text{value A-B}}$$

A second analysis of the ISS sustaining engineering organization looked at potential issues with communications and coordination between functions. Using a similar approach to the critical skills DSM, a communications penalty DSM was generated. Of the 36 activities performed for ISS operations, eight different organizational units are involved and include groups such as Engineering, Safety, Software, and Flight Controllers. A row/column was added to the DSM to denote which one of the eight organizations was responsible for each function. Looking at each cell of the DSM, if the functions dependent

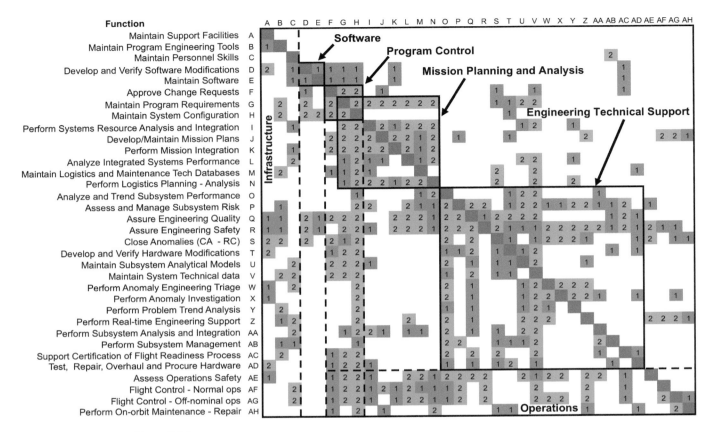

Figure 5.4.2
Organization DSM for ISS sustaining engineering operations.

on each other were performed within the same organization, the dependency value from figure 5.4.2 remained the same. If the two interdependent functions were performed by different organizations, the dependency value in figure 5.4.2 was multiplied by five. The communications penalty between organizational units is represented by the DSM shown in figure 5.4.4.

Results

Assigning attributes to the functions in the organization DSM provides the opportunity to analyze different areas of interest. The critical skills DSM in figure 5.4.3 shows three technical areas with the highest critical skill factors: Mission Planning and Analysis, Engineering Technical Support, and Flight Operations. The DSM identifies specific functions in these technical areas that require technical competency and an in-depth knowledge of

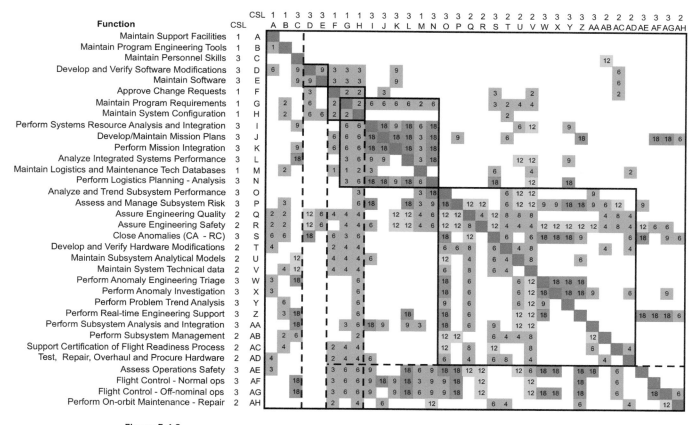

Figure 5.4.3
Critical skills DSM for ISS sustaining engineering operations.

the ISS vehicle. For example, personnel performing logistics planning (row/column N) must understand, in detail, the technical capabilities of critical systems, integrated vehicle performance, and implications for long-term mission planning. All clusters of high critical skill factors were examined. Interviews conducted and review of support contracts showed the ISS personnel in the identified critical areas were highly capable to support long-term ISS operations, and near-term contracts were in place to retain critical skill groups.

Human spaceflight operations are highly complex and involve large teams of people, so the involvement of multiple organizations is not surprising. The communications penalty DSM in figure 5.4.4 highlights specific functions where close coordination across organizational boundaries is required. Eight major organizations support the 36 functions highlighted in the DSM. The large number of high communication penalty interactions helped identify potential areas where cross-organization coordination could pose problems. Review of organizational processes and interviews conducted showed that the use

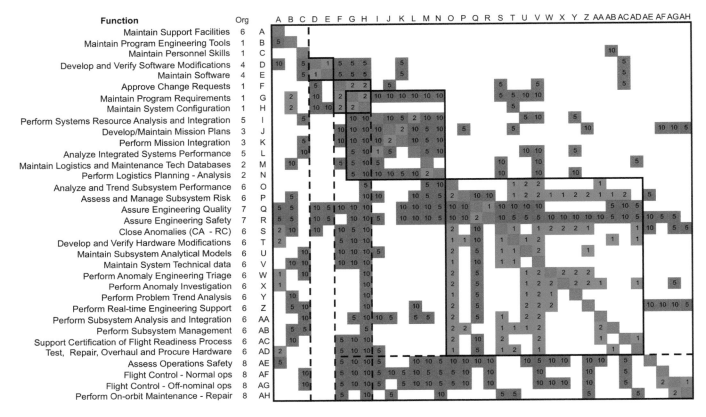

Figure 5.4.4
Communications penalty DSM for ISS sustaining engineering operations.

of formal technical teams and control boards employed by the ISS since 2003 address the need for high levels of coordination. For example:

- Mission Planning and Analysis starts far in advance and is revisited continuously based on the current state of the vehicle. Long-term plans are addressed and coordinated through ISS control boards, and near-term operations are coordinated through a mission management team.

- Major elements of Engineering Technical Support are performed by discipline-specific integrated teams comprised of representatives from all key functions.

- Flight Control Operations is performed by highly trained operators that make real-time decisions. During critical operations, engineering staff are available to provide real-time support to flight controllers. During anomaly resolution, multi-organization problem resolution teams are formed.

Example 5.5   Timken Technology Center

Contributors

Douglas H. Smith
The Timken Company

Steven Eppinger
Massachusetts Institute of Technology

Problem Statement

The Timken Company has been a leading global manufacturer of roller bearings and special alloy steels for many decades. By the late 1990s, Timken was growing rapidly into broader lines of business related to friction management and power transmission. To meet the company's demands for innovation and more effective product development, Timken sought closer ties among its business development, technology development, product design, and manufacturing development functions. It is clear that spatial and organizational designs interact to determine the effectiveness of communications leading to innovation (Allen and Henn 2007). Realizing this, senior management decided it was wise to co-locate these functions at the company's Technology Center located in North Canton, Ohio (figure 5.5.1). To create the most effective organization through co-location, Timken

Figure 5.5.1
Timken Technology Center (courtesy of the Timken Company).

managers wanted to find the best possible spatial layout of the dozens of organizational units relating to new product and business development that were to be housed in the Technology Center. The organization architecture DSM provided a way to analyze interaction data in support of these organization layout decisions.

Data Collection

Timken managers worked with Professor Eppinger to develop a survey for assessment of coordination needs across numerous groups involved in product and business development for the global bearings business. Survey responses were rated on a four-level scale: critical, important, incidental, or none. These interaction data were placed into an organization architecture DSM.

Model

The DSM shown in figure 5.5.2 represents the importance of regular interactions between the 34 organizational units surveyed. The data show that interactions across these units are dense, with some functions such as sales and business development having critical interactions with many others. This level of density makes automated DSM clustering difficult. We therefore used a manual clustering approach, allowing us to manipulate arrangement of the DSM to suggest various possible groupings. Through discussions with Timken managers, we developed the clustering result shown in figure 5.5.2.

Results

The organization DSM shows several clusters of interactions rated critical or important. Moreover, each of these clusters represented groups of people who for years had been spread over multiple locations in the Canton area. For example, the original Technology Center comprised primarily R&D staff. Sales, application support, and business development functions were located in two other buildings and seldom visited the Technology Center. A new physical arrangement of these organizations should address these issues.

The DSM shows which functional units were to be located in the Technology Center, which ones would move to the Corporate Headquarters, and which units would be at globally distributed locations. Of special note in the DSM is the large grouping of core functions, labeled Core Development, Business Development Core, and Manufacturing Side of Core. As noted in the DSM, many of these units have substantial interactions with sales functions located at another office building; however, not everyone could fit into the Technology Center, so we decided to draw the line there.

Implementing the clusters suggested by the DSM analysis would be a major transformation. Nevertheless, managers decided that this was an opportunity to make a change,

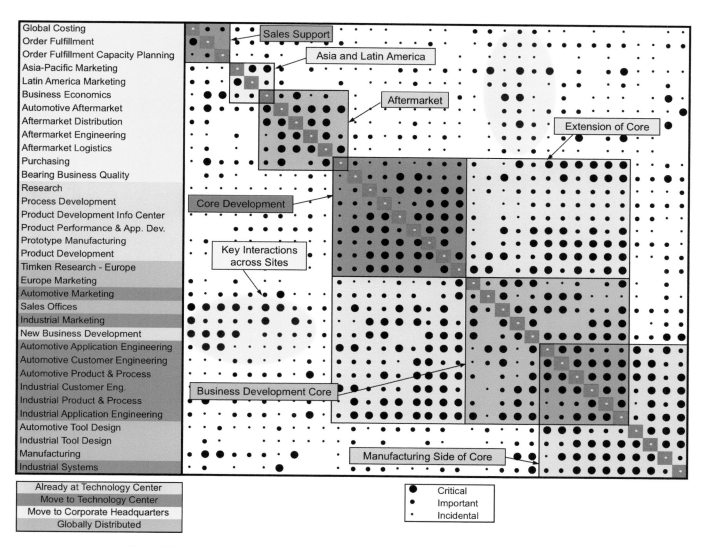

Figure 5.5.2
Organization DSM showing the surveyed importance of interactions across groups related to design and development of new bearings products at Timken.

and Timken implemented recommendations based on the DSM organizational layout in the Technology Center in 1999. This brought key business development, engineering design, and application engineering functions into the same building. For the first time, the physical architecture and the organizational architecture were designed based on the needs for collaboration demanded by the business imperatives.

A senior manager noted, "The analysis showed where organizational affinities were not leveraged in the prior layout. We needed to become a more innovative enterprise, and this showed a path that would help Timken to do just that." In fact, many important innovations came out of the new Technology Center, no doubt in part due to the innovative environment it created and the vigorous engagement of key parts of the organization. In subsequent years, Timken has leveraged the DSM approach for reorganizations and resource deployment studies on several occasions. Its current challenge is to replicate this success at its other global R&D locations.

Reference

Allen, Thomas J., and Gunter W. Henn. 2007. *The Organization and Architecture of Innovation: Managing the Flow of Technology*. Burlington, MA: Elsevier.

adopted this structure with confidence that the new structure would deliver greater project results than the original organization structure.

As shown in figure 5.6.2, the new Government Relations team had substantial overlap of responsibility with the Technical team. Extensive discussions were necessary to decide how to handle the many important interfaces between these teams. It was decided to assign both of these teams' leaders with accountability for the joint deliverables involving both teams but with one of the team leaders (on the Technical team) responsible for managing the interfaces.

Despite the reorganization and the focus given to the complex set of deliverables required to apply for government approvals, the project did not receive the necessary approvals and, unfortunately, was terminated before engineering was completed.

Example 5.7   BP Stakeholder Value Network

Contributors

Wen Feng, Edward F. Crawley, and Olivier L. de Weck
Massachusetts Institute of Technology

René Keller, Jijun Lin, and Bob Robinson
BP p.l.c.

Problem Statement

BP had secured the rights to a significant oil reservoir in a foreign country by creating a multibillion dollar joint venture with a local corporation. Although this multinational energy project would be technically challenging, there were early indications that the complexity of stakeholder relationships would pose a significant risk. In an effort to support the project in understanding these complex stakeholder relationships, we utilized a specialized type of organization architecture DSM to answer the following questions: What are the primary paths for a project to engage stakeholders? Who are the most important stakeholders for the project?

Data Collection

A stakeholder value network is a multirelational network consisting of a focal organization, the focal organization's stakeholders, and the tangible and intangible value exchanges between the focal organization and its stakeholders, as well as between the stakeholders themselves (Feng and Crawley 2008). To understand the impacts of both direct and indirect relationships between stakeholders (including the focal organization), qualitative and quantitative models were built to populate the stakeholder value network. Correspondingly, there were two phases for data collection. First, we surveyed the relevant documents for the project and interviewed the project managers to identify major stakeholders of the project and their roles, objectives, and specific needs. These were mapped as value flows between stakeholders and then taken as the inputs for the qualitative stakeholder model. Second, we designed a questionnaire to ask the representatives of each stakeholder to characterize their specific needs from two aspects: "recipient's intensity of need" and "source's importance in fulfilling the need." These were combined into a utility score for each value flow and then taken as the inputs for the quantitative stakeholder model.

Model

Figure 5.7.1 shows a map visualizing the qualitative model of the stakeholder value network of this multinational energy project, which includes 27 value flows between 9 stakeholders. This qualitative model can also be represented by the left DSM in figure 5.7.2, showing the number of value flows from column stakeholders to row stakeholders. Further, the right DSM in figure 5.7.2 shows the total utility score of value flows from column stakeholders to row stakeholders, which is calculated from the stakeholder questionnaire and provides the inputs for the quantitative model.

Based on the qualitative model and the numerical inputs from the questionnaire, a specific algorithm of DSM multiplication was designed to search all the value paths between any two stakeholders (see figure 5.7.3), which were the basis for the quantitative

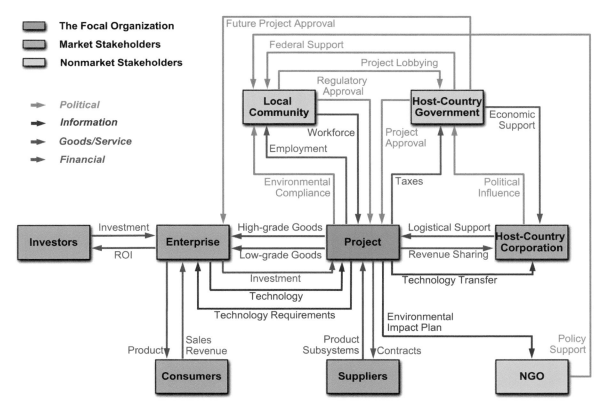

Figure 5.7.1
Stakeholder map for the multinational energy project.

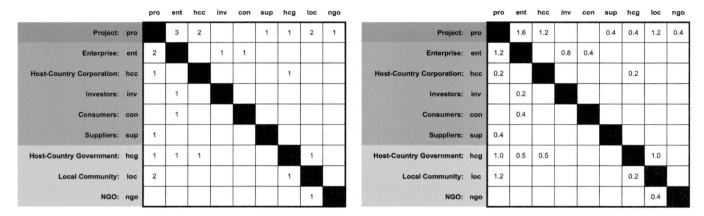

		pro	ent	hcc	inv	con	sup	hcg	loc	ngo
Project:	pro		3	2			1	1	2	1
Enterprise:	ent	2			1	1				
Host-Country Corporation:	hcc	1					1			
Investors:	inv		1							
Consumers:	con		1							
Suppliers:	sup	1								
Host-Country Government:	hcg	1	1	1					1	
Local Community:	loc	2				1				
NGO:	ngo							1		

		pro	ent	hcc	inv	con	sup	hcg	loc	ngo
Project:	pro		1.6	1.2			0.4	0.4	1.2	0.4
Enterprise:	ent	1.2			0.8	0.4				
Host-Country Corporation:	hcc	0.2					0.2			
Investors:	inv		0.2							
Consumers:	con		0.4							
Suppliers:	sup	0.4								
Host-Country Government:	hcg	1.0	0.5	0.5					1.0	
Local Community:	loc	1.2				0.2				
NGO:	ngo							0.4		

Figure 5.7.2
DSM for the qualitative model and the inputs for the quantitative model.

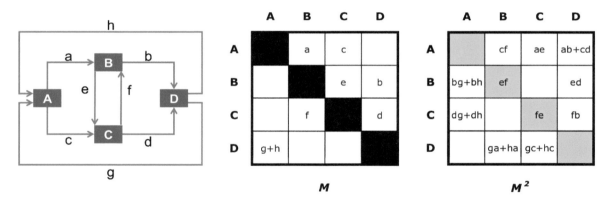

Figure 5.7.3
Example of DSM multiplication to compute reachability.

model. First, a typical stakeholder value network, or a multidigraph, can be represented by a DSM using the addition operation to connect the names of multiple flows between the same pair of stakeholders (see M). Second, multiplying the original DSM by itself once computes a new square matrix, in which the element (i, j) in the resulting matrix represents all the paths from Stakeholder i to j with path length equal to 2 (see M^2). (This result is generalizable and known in graph theory as the reachability or visibility matrix. It is discussed further in chapter 6.) We had generous assistance from Yuan Mei of MIT's Computer Science and Artificial Intelligence Laboratory, who helped in optimizing and implementing the DSM multiplication analysis.

		pro	ent	hcc	inv	con	sup	hcg	loc	ngo
Project:	pro	43	9	6	9	9	1	6	6	1
Enterprise:	ent	2	20	12	1	1	2	12	12	2
Host-Country Corporation:	hcc	6	17	17	17	17	6	5	14	6
Investors:	inv	2	1	12	1	1	2	12	12	2
Consumers:	con	2	1	12	1	1	2	12	12	2
Suppliers:	sup	1	9	6	9	9	1	6	6	1
Host-Country Government:	hcg	6	13	11	13	13	6	30	13	6
Local Community:	loc	6	19	13	19	19	6	7	25	6
NGO:	ngo	6	19	13	19	19	6	7	1	6

		pro	ent	hcc	inv	con	sup	hcg	loc	ngo
Project:	pro	6.2	2.1	1.5	1.6	0.8	0.4	0.9	2.0	0.4
Enterprise:	ent	1.2	2.8	1.8	0.8	0.4	0.5	1.1	2.4	0.5
Host-Country Corporation:	hcc	0.8	1.2	1.1	1.0	0.5	0.3	0.3	1.0	0.3
Investors:	inv	0.2	0.2	0.4	0.2	0.1	0.1	0.2	0.5	0.1
Consumers:	con	0.5	0.4	0.7	0.3	0.2	0.2	0.4	1.0	0.2
Suppliers:	sup	0.4	0.8	0.6	0.7	0.3	0.2	0.4	0.8	0.2
Host-Country Government:	hcg	2.9	4.2	3.9	3.3	1.7	1.2	2.6	3.3	1.2
Local Community:	loc	1.5	2.8	2.2	2.2	1.1	0.6	1.0	3.1	0.6
NGO:	ngo	0.6	1.1	0.9	0.9	0.4	0.2	0.4	0.4	0.2

Figure 5.7.4
DSM for the quantitative model.

Using this algorithm, figure 5.7.4 shows the number of value paths (left) and the total utility score (right) of those paths between any two stakeholders in the multinational energy project. The diagonal elements, or all the value paths beginning from and ending with the same stakeholder, were further analyzed to study the implications of the network for that stakeholder (in this case, we were most interested in the focal organization, the Project).

Results

The following results were obtained from the numerical analysis of the stakeholder value network presented in figure 5.7.1. The first result was a list of primary paths for the Project to engage its stakeholders, which were ranked by the path scores. Figure 5.7.5 highlights the top six paths with a length greater than two steps. These indirect paths are useful for the Project to formulate high-leverage strategies when it is difficult to engage a stakeholder directly. For example, if the Local Community is reluctant to issue the Regulatory Approval, the Project can turn over Taxes to the Host-Country Government and then use the Federal Support from the Host-Country Government to influence the Local Community, as shown in the first path. In fact, project managers confirmed the significance of these paths with real experience. However, without the stakeholder value network analysis, supported by the DSM modeling platform, there is no rigorous way to identify these valuable indirect paths quickly, especially when the size of the network becomes large.

The second result was a ranking of the relative importance of stakeholders for the Project, measured by the Weighted Stakeholder Occurrence (WSO):

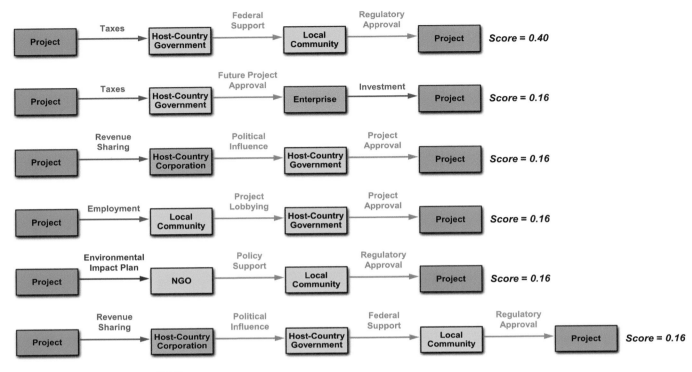

Figure 5.7.5
Project's top six indirect paths.

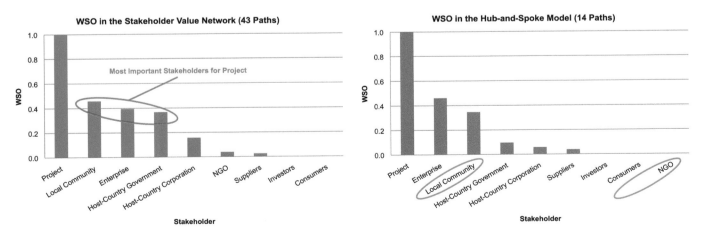

Figure 5.7.6
WSO in the stakeholder value network and WSO in the hub-and-spoke model.

$$\textit{Weighted Stakeholder Occurrence (WSO)} =$$
$$\frac{\textit{Score Sum of the Value Paths Containing a Specific Stakeholder}}{\textit{Score Sum of All the Value Paths for the Focal Organization}}$$

Figure 5.7.6 compares the WSO calculated in the network model with the WSO calculated in the hub-and-spoke model, where only the direct relationships between the Project and its immediate stakeholders are examined. The higher importance of the Local Community and the NGO in the network model was confirmed by managers and historical facts and essentially meant that project teams—when only considering direct relationships with stakeholders—are likely to underestimate the influence of some stakeholders.

References

Cameron, B. G. 2007, June. *Value Network Modeling: A Quantitative Method for Comparing Benefit across Exploration Architectures*. Master's thesis, Massachusetts Institute of Technology, Cambridge, MA.

Feng, W., E. F. Crawley, O. L. de Weck, R. Keller, and B. Robinson. 2010 July. *Dependency Structure Matrix Modelling for Stakeholder Value Networks*. Proceedings of the 12th International DSM Conference, Cambridge, UK.

Feng, W., and E. F. Crawley. 2008. *Stakeholder Value Network Analysis for Large Oil and Gas Projects*. BP-MIT Research Report.

Sutherland, T. A. 2009, June. *Stakeholder Value Network Analysis for Space-Based Earth Observation*. Master's thesis, Massachusetts Institute of Technology, Cambridge, MA.

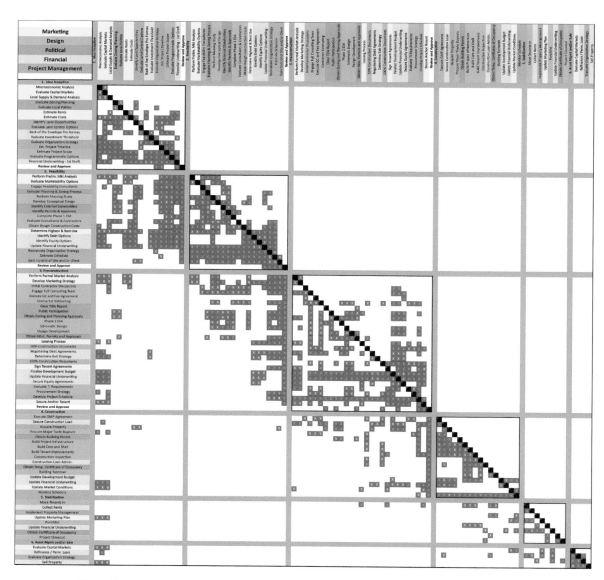

Figure 6.1
A process architecture DSM model depicting a real estate development process (example 7.4).

In this chapter, we consider the architecture of processes, with particular attention to product development processes for engineered systems. We show how DSM is applied to represent and analyze such processes and the types of insights gained through these DSM applications. We begin with a brief synopsis of terminology used in the particular context of process architecture DSM modeling.

Terminology

Process   A system of activities and their interactions comprising a project or business function, such as an engineering design and development project.

Process Architecture   The structure of a process—embodied in its activities and their interactions with each other and the process environment—and the principles guiding its design and evolution.

Activities   The elements of action comprising a process, which in various contexts may be tasks to execute, information to generate, decisions to make, or design parameters to determine. Each activity transforms one or more inputs into one or more outputs. Complex processes are generally broken into phases, stages, or subprocesses, which are further decomposed into activities.

Interactions   The output-to-input relationships between activities. We are especially interested in work products, deliverables, and information flows, where the outputs of activities enable the execution of others.

Process Architecture DSM   A mapping of the network of interactions among the activities in the process, also known as **process DSM**, **process flow DSM**, **activity-based DSM**, and **task-based DSM**.

Sequencing   Analysis of the process architecture DSM through logical ordering of the activities, identifying sequential, parallel, and coupled sets of activities; also known as **partitioning** analysis for process DSM models.

Coupled Activities   A set of two or more activities whose interactions create the potential for iterations, as there exists a (direct or indirect) path of interactions from each activity in the set to every other and back to itself; also known as **feedback loops**, **cycles**, or **circuits** and in graph theory as strongly connected components, vertices, or nodes.

Block   A group of coupled activities identified in the process architecture DSM.

Iteration   The repetition of activities, also known as **rework**. Iterations may be planned (due to coupling or uncertainty) or unplanned (due to discovery of errors or arrival of new information).

Tearing   Analysis of a coupled block of activities to identify interactions for temporary removal from the block, after which the block is resequenced to suggest a better process. Torn marks are reinserted into the DSM and become assumptions or starting points when executing the process—hopefully with minimal iteration.

Terminology
(continued)

> **IR/FAD** Abbreviation for the main DSM convention used in this book, where the off-diagonal marks are oriented as inputs in rows (IR) and outputs in columns. With temporal (process) DSMs, this results in any feedback marks appearing above the diagonal in the matrix (feedback above diagonal [FAD]).
>
> **IC/FBD** Abbreviation for the DSM convention with inputs in columns (IC) and feedback marks below the diagonal (feedback below diagonal [FBD]). An IC/FBD DSM is the transpose of an IR/FAD DSM. The two conventions convey equivalent information. Both are used because each offers advantages.

Background

The disciplines of project management and operations management are largely concerned with the management of process flows (in projects and operations, respectively). Numerous methods are used to plan and schedule the start and end of activities and to coordinate the flow of information (as well as materials, funds, and other transfers) in processes. The most commonly used methods are process flow diagrams or flowcharts—comprised of boxes representing the activities and arrows representing the flow or transfer of information and materials between them—and Gantt charts (Gantt 1919) comprised of bars representing the activities and, sometimes, arrows showing the dependencies as well. Over the past 30 years, DSM has also been applied to processes, yielding a highly useful and potentially richer model of the process architecture and leading to improved process performance. DSM is especially useful when processes are complex and iterative.

The type of DSM used for process modeling is the *process architecture DSM*, also called the *process DSM*, *process flow DSM*, *activity-based DSM*, or *task-based DSM*. This kind of DSM represents the network of activities comprising a process and its interactions. Two variants of the process architecture DSM are the *parameter-based DSM* and the *software process DSM*. In the parameter DSM, the network of design parameter decisions is modeled as a set of activities, each of which determines one or more design parameters (see examples 7.13 and 7.14). In the software DSM, the software process flow is modeled to depict the sequence of execution of the software code (see example 7.15).

The process DSM began as the original DSM technique developed by Don Steward in the 1960s. As noted in chapter 1, Steward was using matrix-based techniques to solve systems of equations, where a key consideration is the order in which the variables should be solved, so as to minimize the need for iteration in the solution algorithm (Steward 1962, 1965). He also realized the applicability of this approach to representing and improving the order of activities in processes. While Steward's work gained only limited circulation in the 1960s and 1970s, others were using square precedence matrices to concisely represent activity sequences (e.g., Fernando 1969; Hayes 1969), and still others applied the algorithms for minimizing cycles in matrix representations of systems (e.g., Warfield 1973). Steward's work on the DSM was finally published in 1981, but it was not until the early 1990s that researchers utilized the DSM methods in earnest. At that time, an explosion of works appeared, mainly by researchers at MIT, applying and extending the original DSM methods. Process DSM applications at NASA (Rogers 1989, 1996), Boeing (Grose 1994), and General Motors (Black et al. 1990; Eppinger et al. 1990) in the 1990s were among the first demonstrations of DSM applied to industrial problems. Since that time, the use of DSMs to model and analyze process architectures has continued to expand, making it the largest area of DSM research and application.

Process modeling is a common and long-established field that uses a variety of methods for modeling and representation (Browning et al. 2006; Browning & Ramasesh 2007). Our work in this area, both in academia and industry, has shown us that a few key points deserve mention regarding processes and process modeling. First, processes exist—whether we model them or not. Every enterprise has processes (a way to get results) even if they are not documented, consistent, effective, or efficient. Process modeling often follows an inductive approach in an effort to document the "as is" reality of how work is accomplished and results are produced. Tremendous value can come from the discoveries made during the building of a process model regardless of any further value derived from analyzing the model.

For clarity in our discussion of processes and process models, we established the definitions at the beginning of this chapter. Three important notes are in order regarding them. First, the terms *process* and *activity* are usually observer-dependent (i.e., one person's process may be another person's activity). This occurs because processes are a kind of system and, as such, exhibit the general property that every system is part of a larger system, and every component of a system may be further decomposed into smaller components. Hence, we use the terms *process* and *activity* in a relative sense, typically using the term *process* to refer to an entire DSM model and the term *activity* to refer to one of the elements within it. Second, it is important to note that many of the work products in processes are just information and may be transmitted informally, meaning that the modeler may need to do additional work to capture these types of interactions. Third, each activity is both a customer/receiver/user and a producer/supplier/provider of work products. That is, each activity both requires input(s) and produces output(s). Similarly, each work product is both an output and an input depending on whether one takes the point of view of its provider or its user. Some activities may have external inputs and/or produce external outputs, which may be captured in an extension to the DSM (see example 7.6).

The architecture of a process has to do with the way its activities work together to deliver results. Specifically, process architecture consists of three types of mappings: (1) hierarchical decomposition of the process into activities, (2) input–output relationships between activities, and (3) various mappings of meta-relationships between activities (such as mutual resource dependencies or multiple instances of similar activities). The

DSM has been applied mainly to the second of these, which calls for a square matrix, whereas the first calls for a work breakdown structure (WBS) and the third usually requires advanced object-oriented modeling and database referencing techniques. Many entire books have been written about process modeling methods. Here we limit our focus to how DSM can be used in process planning and improvement endeavors. (See Browning [2009] for a perspective on how DSM can be used in conjunction with other process model views and object-oriented models.)

Casting Rechtin's (1991) insightful explanation of system architecting in the specific terms of processes (substituting process and activity for system and element, respectively), we get:

- Relationships among [activities] are what give [processes] their added value.
- The greatest leverage in [process] architecting is at the interfaces.

In other words, while many process models emphasize the activities, the interactions among activities play a tremendously important role in the process' ability to deliver value. In fact, the same set of activities may or may not provide value depending on the inputs they use and how they interact. This point stands as an interesting contrast to some of the literature on lean processes, where modelers endeavor to categorize activities as value-adding or non-value-adding according to their intrinsic properties only. However, providing bad inputs to a value-adding activity yields bad outputs (Browning 2003; Browning & Heath 2009).

One of the advantages of the DSM is its emphasis on interactions. Most of the square matrix is devoted to representing the presence (and sometimes various properties) of the interactions, and DSM analysis highlights important patterns of interactions and their implications for process behavior. Moreover, the methods we present below for building a process DSM, which focus on drawing out the flow of information and work products, tend to uncover a relatively rich set of interactions. In contrast, many other process modeling methods and representations that are able to represent interactions will, because of the way they are built and displayed, nevertheless under-represent them. For example, many flowcharts show only a minimal set of arrows between the boxes—just enough to connect them—rather than the full set of inputs and outputs for each activity. Similarly, many Gantt charts do not explicitly indicate the flow of information and work products that establishes activity dependencies.

Figure 6.2 illustrates how the process DSM is used to represent interactions among activities, including these four fundamental types of relationships:

- **Sequential activities**   Output of the upstream activities enables execution of the downstream activities, so they are executed sequentially. Some sequential activities may be partially overlapped by starting the downstream activity before the upstream activity is completed. Overlapping normally sequential tasks to accelerate a process may be achieved by careful scrutiny of each finish-to-start dependency (Krishnan et al. 1997).

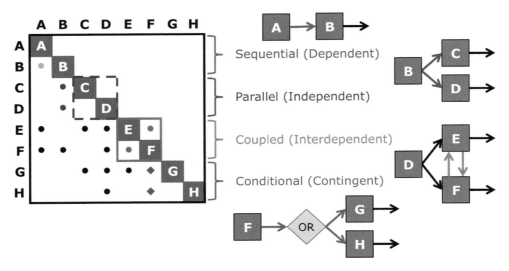

Figure 6.2
Sequential, parallel, coupled, and conditional activity relationships in the process DSM.

- **Parallel activities** Without input–output interaction between them, they may be executed simultaneously. Note that although parallel activities may not have any direct interaction, they might depend on the same resource, from which perspective they do indeed interact. In general, resource constraints are considered in project scheduling only after the primary input–output dependencies have been addressed.

- **Coupled activities** Each needs input from one or more of the others, so they must iterate until they converge on a mutually satisfactory solution. Coupled activities are common in most types of engineering design and development projects, particularly where uncertainties are addressed through invention, analysis, prototyping, verification, validation, and testing tasks.

- **Conditional activities** Execution of the downstream activities is contingent on decisions made in the upstream activity. Although it is uncommon to show contingent process flows explicitly in process DSM models, they may be represented in various ways, such as by using different symbols, as shown with diamond marks in figure 6.2. (This is similar to the use of diamonds in flowcharts to represent decision points.)

Although each of these types of interactions *could* be represented in a flowchart or other process modeling representation, certain patterns—especially the subsets of coupled activities—often go undocumented and unnoticed in such diagrams, whereas the DSM can highlight them. In fact, the most common process modeling and analysis tools—the Project Evaluation and Review Technique (PERT) flowchart, the Critical Path Method (CPM), and their associated Gantt charts—do not well represent cyclical processes nor do they show coupled groups of activities. Coupled tasks would create a rework circuit in a PERT or CPM diagram, but this is not allowed because the method cannot compute the critical path through a process containing a cycle. Some other process modeling techniques, such as value stream mapping and IDEF methods, could include coupled activities, but they are generally not supported by the analysis to identify such cycles.

Process Iterations

In our studies of process architecture, using DSM models of dozens of engineering projects, we have found that process iteration is one of the most salient phenomena stemming from the patterns of interaction in a process. Iteration involves the repetition (or rework) of activities, represented by *feedback loops* or *cycles* in the process. Because these feedbacks are often destabilizing and unplanned, iteration and rework are major drivers of project cost and schedule overruns and associated risks (e.g., Cooper 1993). For example, in a study of nine projects at Intel, iteration accounted for 13% to 70% of project duration, with a mean of 30% (Osborne 1993). (Example 7.2 describes one of these projects.) A model of a preliminary design process at Boeing (example 7.6) showed how the overall duration and cost of a project can change dramatically with changes to the process architecture even without changes to the individual activities themselves (Browning and Eppinger 2002).

Several authors have discussed design as an iterative process (e.g., Kline 1985) and have explored the sources of design iterations (e.g., Eppinger et al. 1994; Steward 1981). Lévárdy and Browning (2009) reviewed the following causes of iterations, stating them in the context of process information flow:

- **Inherent coupling** Activities are structurally interdependent and cannot be executed without assuming, exchanging, checking, and updating information in an iterative fashion.

- **Poor activity sequencing** Information is created at the wrong time (often too late), which forces other activities to wait or make assumptions.

- **Incomplete activities** Information needed by later activities is not fully available, even though the earlier activities have started.

- **Poor communication** Information is not transmitted clearly, promptly, or appropriately.

- **Input changes** External information (or proxy assumptions) used by activities to do their work is subsequently changed (e.g., requirements changes), necessitating rework of those activities and potentially many others that have followed.

- **Mistakes** Defective information is inadvertently created and later discovered to be erroneous, causing rework of portions of the process; a greater time lag until this discovery amplifies the effect (Cooper 1993).

In general, iteration occurs when the cumulative output from prior activities, plus the assumptions that can be reasonably made at the time, are insufficient to enable the next activities to be properly performed to add appropriate value to a project.

Some of these causes of iteration are avoidable through careful process analysis and risk management. Other types of iteration are more fundamental to the process and need to be planned and managed differently. Indeed, some types of iteration should even be encouraged and facilitated so they will converge more quickly (see example 7.6 regarding desired iterations, and see example 7.2 for an illustration and discussion of the distinction between planned and unplanned iterations).

The phenomenon of iteration shows that process architecture matters. The understanding that process architectures can be designed in superior ways, and that superior architectures can provide competitive advantages, has motivated research and development in several areas, including that which uses the DSM. Some of these advantages include: minimizing unplanned rework and iterations, negotiating input–output relationships and commitments, reducing project duration and cost, and reducing the risks associated with not meeting deadlines and budgets. Of course, these are in addition to the general advantages for complexity management, visualization, understanding, and innovation mentioned in chapter 1.

Building a Process Architecture DSM

The basic procedure for building a process architecture DSM is as follows:

1. Decompose the overall process into its activities (via intermediate subprocesses and phases/stages if needed). Lay out the square DSM with activities labeling the rows and columns, listed in the usual sequence (if known) and grouped into subprocesses or phases/stages if appropriate.

2. Identify the known interactions (input–output relationships) between the activities and represent these using marks or values in the DSM cells.

The process employed at Boeing for the conceptual design of an unmanned combat aerial vehicle (UCAV) illustrates the procedure for construction of a process architecture DSM model. The UCAV DSM is shown in figure 6.3, with both the original process sequence (top DSM) and an alternative ordering of the activities (bottom DSM), which is discussed later in this chapter. The first activity sets the design requirements and objectives (DR&O), and the second activity suggests a configuration concept to meet them. Activities 3–10 then analyze and evaluate the concept from various discipline perspectives, exchanging information and data as they go. Activity 11 collects all of these results and decides whether to proceed to activity 12—which wraps up the phase and prepares for the next one (preliminary design)—or to iterate by changing the design concept, the

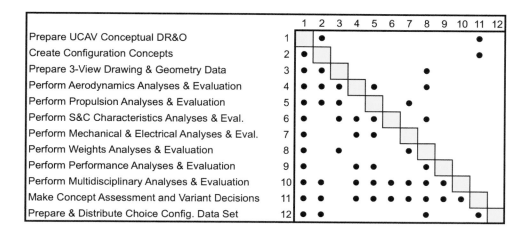

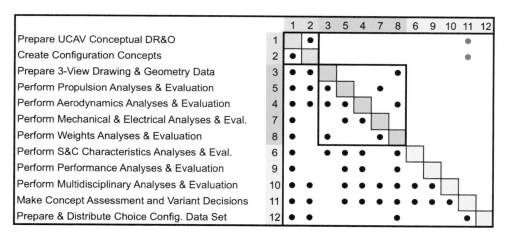

Figure 6.3
Process DSM model of the UCAV conceptual design process (example 7.6) before sequencing (top) and after sequencing (bottom).

DR&O, or both (as represented by the two marks in the upper-right corner of the DSM) (see example 7.6 for further details and discussion).

Our experience building process DSM models has taught us many lessons about data collection and graphical representation. These lessons have been helpful in a variety of DSM applications:

- **Sequencing**   Established processes have a typical sequence, usually documented in a process flow diagram, Gantt chart, or list of activity start dates or due dates. This typical sequence is usually a good starting point for the initial activity list used in the DSM model. Subsequent analysis may scrutinize or optimize the activity sequence, revealing insights about performance and suggestions for process improvement.

- **Process decomposition**   Established processes also often have an established decomposition of activities (e.g., a WBS). This decomposition provides a good starting point for building the model, even if it must be refined later. Regardless, it is important to ensure that the model's constituent activities indeed capture all of the work done to execute the process. Sometimes seemingly minor activities can have a significant effect on a process, so it is important to include them in the model, even if only as part of another activity. Another key issue may also arise, especially for large, multidisciplinary development processes, and that is which way to do the decomposition. Should the process be broken down first by discipline or by a particular product subsystem, forming subprocesses that span the overall process' entire duration? Or should the process be parsed first by stages or phases, each containing all activities but only for a portion of the overall process' duration? Although either approach should eventually reach the same individual activities, debating the options can bog down a modeling effort. The most important thing is just to pick *an* approach, making modifications later if needed.

- **Convention for representing input, output, and feedback**   There are two conventions for orienting process DSM models. The more common representation for DSMs (originating from Steward) places activity inputs in the rows (IR) and activity outputs in the columns, resulting in *feedback above the diagonal* (IR/FAD). We have used this convention throughout this chapter. An alternative convention (originating from N^2 and IDEF0 diagrams) places inputs in the columns (IC) and outputs in the rows, resulting in *feedback below the diagonal* (IC/FBD). Four examples (5.2, 7.6, 9.7, 9.11) show DSM models using the IC/FBD convention. We have heard lengthy debates over the merits of each convention, but no absolute standard has yet emerged. We have concluded this is mostly a matter of personal preference based on familiarity with reading one type or the other, although the IC/FBD convention does provide a benefit with respect to the orientation of external input and output regions around the square DSM, as demonstrated in example 7.6. However, it is important to stress that the two conventions convey equivalent information; each is just the matrix transpose of the other.

- **Building and verifying the model**   In principle, it is possible to create a process DSM by filling out either the rows or the columns. In the IR/FAD convention, filling the rows means identifying the inputs for each activity and placing these marks in the appropriate columns depending on their source. Filling the columns would mean listing where each activity's outputs go. We have found that it is generally more reliable to ask process owners where their inputs come from. They know what they need because they generally have to seek out their input information. However, they may not reliably know where their outputs are used. Of course, it is also a good practice to discuss outputs. An eye-opening approach is to build two DSMs—one by rows (i.e., with data on activity inputs) and one by columns (with data on activity outputs)—and compare these to verify the model.

- **Modeling the as-is process first**   Experienced process modelers know that it may not be easy to capture the actual process. People will often explain how the process *should be* rather than describing the process *as is*. In fact, each of these process variants has a distinct meaning. We generally recommend to capture the as-is process first and then use insights about what process owners feel should be different to work toward process improvements. Jumping to supposed improvements without a valid baseline process model for comparison and discussion can lead to unexpected results.

- **Accounting for process iterations**   Documenting process iterations can be difficult because many iterations represent errors (as discussed earlier) and many people have been taught to think of all rework as wasteful. We find it helpful to begin by discussing the established process—as planned. Then discuss the exceptions—how the process can fail to go as planned. This discussion may uncover many of the known failure modes of the process. Many of the examples in the following chapter (such as example 7.2) took this approach. A full failure modes and effects analysis (FMEA) can even be done. It can also be difficult to model planned iterations, especially when each iteration involves executing each activity in a different mode using different (or more mature) inputs and producing different (or more mature) outputs. In some cases, it is helpful to "unroll" planned iterations and represent them as a repeated set of activities in the overall process.

- **Interaction strength**   It is often useful to distinguish the strength (or other attributes) of each interaction using numerical values, creating a numerical process DSM model. Several examples in chapter 7 show various ways to quantify interaction strength. For instance, some advanced models use probability and/or impact data for each interaction (e.g., Browning and Eppinger 2002; Smith and Eppinger 1997). Interaction strength can also vary over successive iterations (Eppinger et al. 1997; Lévárdy and Browning 2009).

- **Highlighting coupled activities**   Coupled activities are often identified in the DSM using shaded or outlined square boxes along the diagonal, enclosing the marks coupling the activities (see examples 7.2, 7.5, 7.12). Sometimes parallel activities are shown in a similar way using dashed boxes or alternating shaded bands (Grose 1994).

- **Visualization guidelines**   Use appropriate graphics to allow the DSM model to help explain the process. We have found many ways to use colors, shading, symbols, labels, and other notations to highlight a wide variety of interesting phenomena. The examples in chapter 7 illustrate various uses of graphics to add explanatory power to DSM models.

- **Granularity of the model**   Every process can be modeled at several levels of decomposition. This is primarily a tradeoff of modeling effort versus richness. As the examples in chapter 7 show, many highly insightful process DSM models are decomposed in the range of 30 to 70 activities. While building the model, it is not uncommon to discover specific subprocesses, phases, or activities that merit further decomposition to provide insight into the situation underlying the actions and interactions within. Similarly, it may be discovered that some sections of the model could be aggregated without much loss of insight.

- **Accounting for external inputs and outputs**   External inputs and outputs can be represented in a process DSM by using additional rows and columns. These are usually placed outside of the main matrix using the IC/FBD convention (see examples 7.6 and 7.11).

- **Model boundaries**   A DSM model may represent only a portion of a larger process. This is a useful way to focus on a particularly important (and perhaps complex) portion of a larger process that needs to be better understood or improved. In this case, however, important actions and interactions may reside outside the model's scope, so it may be helpful to include some of the larger (external) process in the model or at least account for the interactions with the external input and output regions around the DSM.

- **Additional attributes of activities and interactions**   Although not usually shown explicitly in a DSM, many DSM analyses utilize additional attributes of the activities (e.g., duration, cost, learning curve, probability, and/or impact of input change) and interactions (e.g., work product requirements, information maturity, and probability of change) (Browning 2009). Depending on the purpose of the model, the modeler should consider the extent to which such further information should be gathered. While adding richness to the model and extending its capabilities, these additional data will also increase the time required to build the model.

- **Validating the model**   It is important to have process owners and workers review and discuss the model. They should scrutinize any initial insights or findings to see whether they could be better explained by a required improvement to the model.

The time required to build a process DSM model depends on the amount of data required and the effort needed to acquire it. As the number of activities grows, the size of the DSM increases quadratically, as does the number of potential interactions, but our experience and some data (Whitney et al. 1999) have shown that this growth is actually more linear. Whitney et al. (1999) examined several DSM models and found about six input marks on average. The time required per activity depends on a number of factors, such as the breadth of input types, the availability of existing documentation to draw from, the latent process knowledge available from contributors, the experience of the modeler, the medium of data gathering (survey, online tool, meetings, etc.), and the purpose of the model. We have encountered a wide range of possible times, with many models falling in the range of 15 to 45 minutes per activity. In any case, additional richness should be added to the model only where justified by its purpose.

Analyzing the Process Architecture DSM

Sequencing

The most common method of analysis applied to process architecture DSM models is called *sequencing*. This is a form of DSM partitioning analysis that involves reordering the rows and columns of the DSM to minimize iterations (cycles) (i.e., to arrange the activities with as many interactions as possible below the diagonal [in the IR/FAD convention]). There are several algorithms for DSM sequencing; some are applied only to binary DSMs, and others are used to sequence numerical DSMs based on the strength of the interactions. Meier et al. (2007) provided a survey of various sequencing algorithms for binary DSMs.

As mentioned in chapter 1, a process architecture DSM includes a temporal dimension. Unlike the static DSM model types (product and organization architecture), where all of the elements exist simultaneously, the activities in a process DSM usually begin and end at different times. Because the value added by an activity depends on its inputs, it is usually preferable to perform the activity when all of its inputs are ready and available. Because each input comes from some other activity (or from an external source), the input–output relationships among the activities provide the initial determinant of their appropriate sequence. When an activity begins without all of its inputs, it must use assumptions as a proxy for the missing inputs. Being able to begin without all inputs, by making assumptions, is a double-edged sword in a project process. This is much harder (if not impossible) in manufacturing processes because an assembly activity cannot occur until all of the component parts are physically present. But in project activities, many of the inputs are information and therefore, for better or worse, can be assumed. Using assumptions adds risk, however—risk that the assumptions will be partially or even completely invalidated when the actual input becomes available. We can think of this risk as the possibility of having to rework some or all of the activity, as well as any other activities that have already relied on that activity's output, all of which adds time and cost to the process.

Thus, the first heuristic for sequencing is to find the order of activities that minimizes the amount of feedback in the process (i.e., the sequence that minimizes the need for activities to use assumptions). (As noted in example 7.6, the fastest processes do not always have a minimal amount of feedback, but it is often a good heuristic.) Hence, the first heuristic entails an objective of minimizing the number of feedback marks in the DSM. A more sophisticated heuristic recognizes that short feedbacks are preferable to long ones. This second objective entails minimizing some combination of the number of feedbacks and their distance from the diagonal, because a mark's distance from the diagonal roughly indicates the scope of the feedback, with a mark in the upper right corner of the DSM (with IR/FAD) indicating a potential return from the end of the process all the way back to the beginning. Such *long feedbacks* are especially problematic because many more activities will have occurred in the interim period between the initial completion of the upstream activity and its rework (caused by the far downstream activity). These interim activities proceeded with what they thought were valid inputs, but that turned out to be errors. When the upstream activity is reworked, however, it is likely that its outputs will change, thus precipitating a cascade of rework through the process. Thus, long feedback loops are usually much worse than short ones. However, it is important to remember that minimizing feedback loops of any kind is still just a proxy for the real objectives, which are minimizing the process duration and cost or, better yet, maximizing the value of the process results (Browning 2003).

Although many process DSM models are sequenced manually, guided by the previous heuristics, several automated sequencing algorithms are available in software tools. Figure 6.4 illustrates a simple method called path searching based on Steward's (1981, p. 54–55) original algorithm, later summarized by Gebala and Eppinger (1991):

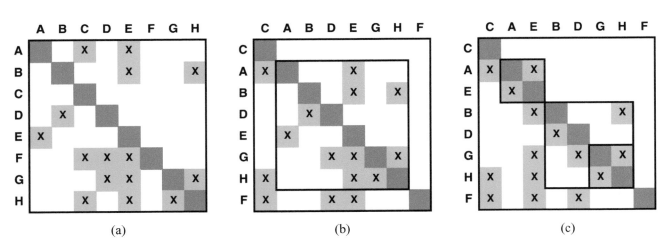

Figure 6.4
Illustration of DSM sequencing based on Steward's path searching algorithm.

1. Sequence the activities that are not part of any cycles (loops). Activities with empty rows have all required information and can be performed first. Activities with empty columns provide no information required by future activities and can be performed last. Once an activity is sequenced, remove it from further consideration. Repeat until no empty rows or columns are found.

2. Identify cycles by one of the three methods mentioned below. Group together all activities in a cycle as a single activity and sequence the group as above if the group has an empty row or column.

3. Repeat steps 1 and 2 until all activities have been sequenced.

Basic sequencing is often called *block diagonalization* or *block triangularization* because it yields a lower triangular matrix (in the IR/FAD convention) where any remaining superdiagonal marks are enclosed in blocks that represent the subsets of coupled activities. Starting with the unsequenced DSM in figure 6.4a, activity C has a blank row (i.e., no inputs from other activities in the process), so it is moved to the beginning, and F has a blank column, so it is moved to the end, yielding the DSM in figure 6.4b. Because every remaining activity has an off-diagonal mark, we arbitrarily begin with activity A to generate a list of successors. This generates the sequence A → E → A. Thus, we isolate A and E as a cycle, and we group them together. We then note that the group AE has no inputs from B, D, G, and H, so we move AE up. Taking the next remaining activity, we find sequence B → D → G → H → B, which implies another cycle. The approach can also be applied recursively to reveal subcycles such as G → H → G. We now have the DSM in figure 6.4c, which contains two main blocks of coupled activities.

A second way to find all of the coupled groups of activities in a DSM model utilizes a linear algebra technique known as the *powers of the adjacency matrix* (Gebala and Eppinger 1991; Ledet and Himmelblau 1970; Warfield 1973). The adjacency matrix is simply the binary version of a DSM (placing ones in the cells with marks and zeros elsewhere). The Boolean square of the adjacency matrix identifies all the indirect connections two steps removed, the Boolean cube of the matrix finds all connections three steps removed, and so on. The powers of the adjacency matrix are useful for determining cycles because any activity in a cycle must be reachable from itself. An activity is reachable from itself in x steps if, in the adjacency matrix raised to the xth Boolean power, the activity has a nonzero entry on the diagonal. This is illustrated in figure 6.5, which takes the submatrix from Figure 6.4b (because activities C and F have already been determined not to reside in cycles). The Boolean square of the adjacency matrix reveals that activities A, E, G, and H are in a two-step cycle. The fourth power of the adjacency matrix reveals that activities B, D, G, and H are in a four-step cycle. The higher powers of the adjacency matrix reveal no additional cycles in the system. From these, we determine that A and E are involved in a two-step cycle, G and H are involved in a separate two-step cycle, and B, D, G, and H are all part of a four-step cycle, of which the G–H cycle is a part—the same

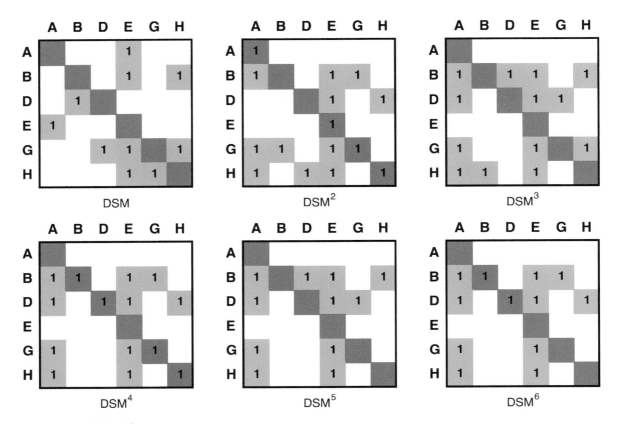

Figure 6.5
Powers of the adjacency matrix for the sub-DSM in figure 6.4b.

structure reached in figure 6.4c. However, finding the blocks of coupled activities by powers of the adjacency matrix is a computationally intensive operation for large matrices.

Note that the N matrices derived by raising the adjacency matrix to successive powers can be overlaid (summed via Boolean arithmetic) to produce the *reachability* or *visibility* matrix, which shows all direct and indirect interactions between the elements (see e.g., Warfield 1973). Figure 6.6 shows a version of the reachability matrix for the DSM in figure 6.4a, where the numbers indicate the number of steps separated in the indirect connection between activities. Because no activity is more than four steps separated from any other, the fifth and sixth powers of the adjacency matrix do not provide any further information (see also examples 3.5 and 5.7).

A third and more efficient way to isolate the subset of coupled activities is to use Tarjan's (1972) depth-first search algorithm. In a manner similar to Steward's path searching

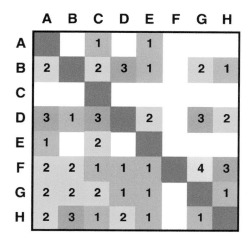

Figure 6.6
Reachability matrix showing all indirect connections, with numbers added to show the number of steps removed.

method, with a linear order of computational complexity (in the number of activities and interactions), Tarjan's depth-first algorithm follows each activity's outputs to determine any dependency paths that cycle back to an activity.

The lower portion of figure 6.3 shows the result of sequencing analysis applied to the UCAV DSM model. This particular binary DSM is quite constrained in terms of sequencing because of the large number of interactions among the activities in the matrix. The two feedback marks from activity 11, shown in red in the upper right of the matrix, are ignored (torn) in the sequencing because they represent planned but optional iterations. Activities 1 and 2 are found to be coupled but have no other inputs, so they remain at the start. However, moving activity 5 upstream moves a mark from above to below the diagonal, and moving activity 6 downstream reduces the size of the second block.

Although sequencing a totally randomly ordered set of activities can show an enormous reduction in the amount of feedback, a basic sequencing analysis may not much alter the original sequence of a working process. However, further analysis can be done to address and manage the remaining feedback loops. We discuss this next.

Coupled Blocks

Once the blocks of coupled activities have been identified, there are several options for dealing with them. In practice, the interdependencies are often just acknowledged, and the individuals and teams executing the activities are told to work together until they have converged on a solution. Many of the integrative mechanisms discussed in chapter

4 apply here (e.g., co-locating the people involved in the coupled activities, utilizing collaboration tools, etc.). One significant benefit of the DSM is that the coupled blocks can be highlighted and the integrative mechanisms can be focused more carefully where they can be most beneficial. However, some groups of coupled activities are linked more strongly than others. Some coupled blocks may be merely an artifact of the model's level of abstraction. In many complex engineering design projects, a substantial portion of the activities may be coupled (e.g., figure 6.3). We therefore present several methods for resolving the blocks.

- **Further decomposition**   One approach is to see whether the coupled block of activities may be decomposed into smaller activities or parameters and then resequenced to reveal a less coupled subprocess. For example, two coupled activities in a DSM are actually aggregations of many smaller activities. When each activity is broken down into its constituents and the interactions are explored more specifically at that level, then a more linear sequence can usually be found among the lower level activities (see examples 7.13 and 7.14).

- **Aggregation**   It is also possible to represent the model at a higher level of abstraction by reducing a coupled block to appear as a single activity, thereby hiding the feedback marks. This approach is not often recommended, however, because it essentially sweeps the issues of interest under the rug, so to speak.

- **Adding new activities**   New activities may benefit the process by creating information at a different point (e.g., earlier), thereby allowing other activities to use real information instead of making assumptions that may cause rework, or by decoupling the flow between other activities (see examples 7.2 and 7.5, as well as a fuller explanation in Lévárdy and Browning 2009).

- **Tearing**   Tearing is a systematic method of suggesting an effective way to execute a block of coupled activities with minimal iteration. Tearing involves several steps (as illustrated in figure 6.7):

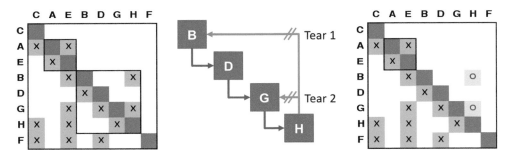

Figure 6.7
DSM tearing analysis identifies marks to remove from coupled blocks in the DSM.

1. Find one or more marks to tear out of the block to reduce the coupling most effectively. Steward (1981) explained how to find tears by drawing the block as a node-link diagram. The best link to tear is the one that breaks the most and the longest circuit(s). For example, see tear 1 in figure 6.7.

2. Suggested tears must be accepted or rejected based on knowledge of the process. Torn marks become assumptions to facilitate execution of the coupled process with minimal iteration. A suggested tear would be accepted if the process owners believe the necessary input can be assumed with some confidence. If the input cannot be assumed, then the next-best tear from step 1 may be suggested.

3. The coupled block is now resequenced by ignoring the torn mark(s). This should break the block into one or more smaller blocks and/or individual activities—eventually a set of fully sequential and parallel activities. In figure 6.7, the first tear leaves a smaller block, which is reduced via the second tear.

4. The torn mark(s) must be replaced in the DSM to serve as a reminder to make the assumption(s) when executing the process and to check the assumption(s) when the activity generating each torn output is executed. (This was also illustrated in figure 6.3.)

5. Because a torn mark represents an assumption, any such assumption that turns out to be invalid will probably subject the process to a rework loop (iteration).

6. Note that multiple tears may be enabled by a single assumption. In figure 6.7, assumption of the result(s) of H may allow both B and G to be executed.

Many process DSM applications use sequencing analysis followed by tearing analysis (see examples 7.1, 7.3, 7.7).

In addition to the traditional DSM analyses described earlier, several advanced techniques have also been developed for analyzing process DSM models, including the following:

- **Simulation**   Discrete-event, Monte Carlo simulation provides a method to predict the distribution of process cost and duration based on a process DSM model augmented with numerical values for the following effects: activity cost and duration, rework probability, rework impact, finish-start overlapping, learning curves, resource constraints, and many other process flow details that can be represented using logic and mathematical expressions (Browning and Eppinger 2002; Cho and Eppinger 2005) (see examples 7.6, 7.10, and 7.12).

- **Eigenstructure**   A powerful analysis applies to the special case of parallel iteration, where coupled activities are executed simultaneously and then exchange information, creating rework as modeled by the DSM interaction values (Smith and Eppinger 1997a). Analysis of the DSM as a work transformation matrix led to the understanding of the phenomena of design convergence and design churn (Yassine et al. 2003) in which iterations may continue while adding little value.

- **Signal flow graphs and reward Markov chains**   The special case of sequential iteration, where coupled activities are executed one at a time based on probabilistic rework, can be analyzed using several types of analytical models. The signal flow graph method borrows a technique used to time signals through circuits (Eppinger et al. 1997). The reward Markov chain method uses numerical values in the DSM to represent rework probabilities and can be used to decide the best execution sequence for a coupled block (Smith and Eppinger 1997b).
- **Meta-heuristics**   Meier et al. (2007) used an enhanced genetic algorithm to sequence binary DSMs. They presented several interesting findings regarding various objective functions for sequencing and the scale-up behavior of solution difficulty as a function of DSM density.

Applying the Process Architecture DSM

Process architecture DSMs have been applied to a range of industrial problems and have produced many useful insights. Many examples are given in the next chapter. Typical applications include:

- Representing and visualizing processes and information flow, which is a common theme through all of the applications presented in chapter 7.
- Analyzing and improving processes, yielding leaner and more streamlined flow. Most of the examples in chapter 7 demonstrate this.
- Planning a project and developing a realistic schedule based on a more detailed process model than is typically used by project management software. For example, a DSM can be converted into a Gantt chart, although this requires some assumptions about the coupled blocks and the likelihood of iterations (see examples 7.1, 7.3, 7.6, 7.7).
- Managing interfaces between process activities, phases, and stages (see examples 7.4, 7.5, 7.8).
- Highlighting iteration and rework (process FMEA) (see examples 7.2 and 7.12).
- Analyzing process cost, schedule, and risk (see examples 7.6 and 7.12).
- Providing an organized framework and/or a graphical user interface for a process knowledge database (Browning 2009; Browning et al. 2006).
- Identifying needs for cross-functional, cross-team interactions (hybrid process-organization DSM) (see examples 7.4 and 7.8).

Conclusion

DSM provides an effective representation for process systems of activities and their interactions. The process architecture DSM can be analyzed via DSM sequencing analysis, which suggests a logical process flow and identifies coupled blocks of activities. Coupled blocks are executed in an iterative fashion and can be further analyzed by several methods, including tearing, which suggests assumptions that can be made to facilitate solution of the coupled activities.

The value of the DSM increases as processes become larger and more complex because such cases make it less likely that any one individual will have an accurate mental model of all the work needing to be done and because they create a need for individuals to communicate, compare, and integrate their partial models of the process.

Process architecture DSM models have been shown to be highly useful because they can:

- generate and represent alternative perspectives on process architecture
- help improve process understanding
- facilitate process innovation
- reduce process cost, duration, and risk

References

This list of references provides additional background on the process architecture DSM.

Don Steward's work on efficient means for solving systems of equations resulted in the term *design structure matrix* and began the application of process DSMs, the identification of blocks of coupled activities, and the breakdown of these circuits via tearing. Warfield built on this work, and Tarjan provided an efficient algorithm for finding the subset of coupled activities.

Steward, Donald V. 1962. On an Approach to Techniques for the Analysis of the Structure of Large Systems of Equations. *SIAM Review* 4 (4):321–342.

Steward, Donald V. 1965. Partitioning and Tearing Systems of Equations. *Journal of the Society for Industrial and Applied Mathematics: Series B, Numerical Analysis* 2 (2):345–365.

Steward, Donald V. 1967. *The Design Structure System*. General Electric internal report no. 67APE6.

Steward, Donald V. 1981a. *Systems Analysis and Management: Structure, Strategy, and Design*. New York: PBI.

Steward, Donald V. 1981b. The Design Structure System: A Method for Managing the Design of Complex Systems. *IEEE Transactions on Engineering Management* 28 (3):71–74.

Denker, Stephen, Donald Steward, and Tyson Browning. 2001. Planning Concurrency and Managing Iteration in Projects. *Project Management Journal* 32 (3):31–38.

Tarjan, Robert. 1972. Depth-First Search and Linear Graph Algorithms. *SIAM Journal on Computing* 1 (2):146–160.

Warfield, John N. 1973. Binary Matrices in System Modeling. *IEEE Transactions on Systems, Man, and Cybernetics* 3 (5):441–449.

In the early 1990s, additional works applied the process DSM to engineering design processes and complex development projects.

Black, Thomas A., Charles F. Fine, and Emanuel M. Sachs. 1990. *A Method for Systems Design Using Precedence Relationships: An Application to Automotive Brake Systems*. MIT Sloan School of Management, Working Paper no. 3208.

Eppinger, Steven D., Daniel E. Whitney, Robert P. Smith, and David A. Gebala. 1994. A Model-Based Method for Organizing Tasks in Product Development. *Research in Engineering Design* 6 (1):1–13.

Gebala, David A., and Steven D. Eppinger. 1991, September 22–25. *Methods for Analyzing Design Procedures*. Proceedings of the ASME International Design Engineering Technical Conferences (Design Theory & Methodology Conference), Miami, FL, pp. 227–233.

Grose, David Lee. 1994, September 7–9. *Reengineering the Aircraft Design Process*. Proceedings of the 5th AIAA/USAF/NASA/ISSMO Symposium on Multidisciplinary Analysis and Optimization, Panama City Beach, FL, Technical Papers, Pt. 1 (A94–36228 12–66).

Apart from the context of DSM, some authors have highlighted the phenomena of iteration and rework in processes and their implications.

Kline, Stephen J. 1985. Innovation Is Not a Linear Process. *Research Management* 28 (2):36–45.

Cooper, Kenneth G. 1993. The Rework Cycle: Benchmarks for the Project Manager. *Project Management Journal* 24 (1):17–21.

Several authors provided advanced techniques for analyzing a process DSM.

Rogers, James L. 1989. *A Knowledge-Based Tool for Multilevel Decomposition of a Complex Design Problem*. NASA Technical Paper no. TP-2903.

Rogers, James L. 1996. *DeMAID/GA User's Guide—Design Manager's Aid for Intelligent Decomposition with a Genetic Algorithm*. NASA Technical Manual no. TM-110241.

Eppinger, Steven D., Murthy V. Nukala, and Daniel E. Whitney. 1997. Generalized Models of Design Iteration Using Signal Flow Graphs. *Research in Engineering Design* 9 (2):112–123.

Smith, Robert P., and Steven D. Eppinger. 1997a. Identifying Controlling Features of Engineering Design Iteration. *Management Science* 43 (3):276–293.

Smith, Robert P., and Steven D. Eppinger. 1997b. A Predictive Model of Sequential Iteration in Engineering Design. *Management Science* 43 (8):1104–1120.

Smith, Robert P., and Steven D. Eppinger. 1998. Deciding Between Sequential and Parallel Tasks in Engineering Design. *Concurrent Engineering: Research and Applications* 6 (1):15–25.

Yassine, Ali A., Nitin Joglekar, Dan Braha, Steven D. Eppinger, and Daniel E. Whitney. 2003. Information Hiding in Product Development: The Design Churn Effect. *Research in Engineering Design* 14 (3):145–161.

Meier, Christoph, Ali A. Yassine, and Tyson R. Browning. 2007. Design Process Sequencing with Competent Genetic Algorithms. *Journal of Mechanical Design* 129 (6):566–585.

Browning's 1998 dissertation led to the first simulation of a process DSM, which provided a foundation for several extensions (see additional references at the end of example 7.6).

Browning, Tyson R., and Steven D. Eppinger. 2002. Modeling the Impact of Process Architecture on Cost and Schedule Risk in Product Development. *IEEE Transactions on Engineering Management* 49 (4):428–442.

Cho, Soo-Haeng, and Steven D. Eppinger. 2005. A Simulation-Based Process Model for Managing Complex Design Projects. *IEEE Transactions on Engineering Management* 52 (3):316–328.

Lévárdy, Viktor, and Tyson R. Browning. 2009. An Adaptive Process Model to Support Product Development Project Management. *IEEE Transactions on Engineering Management* 56 (4):600–620.

Browning's work in the 2000s placed the process DSM in the larger context of process modeling, process improvement, and project management.

Example 7.1   Bioscience Facility at University of Melbourne

Contributors

Elke Scheurmann
Rapid Invention Pty Ltd.

Delyth Samuel
University of Melbourne

Problem Statement

In 2010, one of the bioscience institutes of University of Melbourne wanted to obtain funding for a multistory building extension to their existing facility. This new building annex was to become an integrated science facility and support Australia's biotechnology sector through co-location and enhanced engagement with a major industry company and a high school-level science education facility. This multi-engagement approach was considered important for the future because it was expected to increase interdisciplinary research discoveries for the university and industry and expose more school students to the possibilities of science as a career.

Rapid Invention, a project management consulting firm, was engaged by the university to develop a project plan when the project had already been underway for at least six months. We were asked to translate the multitude of stakeholder requirements into a project plan that could become the basis for a solid funding proposition and lead to a business plan and funding commitments from four major targeted funders (state government, the university, a commercial company, and a major philanthropic trust). The construction phase of the extended facility was to begin within six months so that the building extension could be occupied within 24 months.

We used DSM for planning the project scope from the ground up since the multiple agendas and requirements of the stakeholders and the uncoordinated project planning activities had not uncovered all the unknowns and interdependencies between project elements. We expected a high overall complexity of the project and anticipated the identification of hidden risks that could lead to project failure and delays. A plan to manage any identified risks could then be put into place.

Data Collection

Our first step was to clarify and agree on the aims and objectives of the project with the university, develop a stakeholder list, and then build a high-level project work breakdown structure (WBS) by interviewing relevant individuals in the stakeholder organizations.

The majority of the project tasks initially related to clearly identifying stakeholder requirements and the interdependencies and alignments between them. The second step was to define input and information dependencies between the identified project tasks. We defined the dependencies by three strength levels (low, medium, and high) according to Yassine et al. (1999). This step was followed by an analysis of the automatically created DSM and optimization of the task sequence through partitioning and tearing of dependencies within iteration blocks. We used ProjectDSM 1.0 project planning software (www. ProjectDSM.com), which provides an automated DSM optimization step for the triangulation of the DSM and advice wizards to help users optimize task sequences within coupled blocks.

Model

Although the multitude of stakeholders and requirements made the project appear complex and confusing when we started, the DSM resulted in a straightforward task sequence with only four coupled blocks, as shown in figure 7.1.1.

The four coupled blocks (in sequence) related to:

- Requirements relating to collaborative working relationships and facilities among the various stakeholder organizations.
- Utility requirements of the various co-locating groups.
- Interdependencies among utility vehicle, staff vehicle (cars, bicycle, and motorbikes) and pedestrian access, walk and driveways, loading ramps and docks, safety, and parking, with impact on building design and costing.
- Interdependencies between those making funding commitments.

Two tearing steps simplified the largest (third) coupled block by making two assumptions about pedestrian security and access and giving priority to utility vehicle access over staff vehicle assess due to the physical location details near a busy city road. The resulting process DSM model is shown in figure 7.1.2.

Results

Prior to our involvement, the project had been suffering from an ad hoc planning process within the lead organization (the existing bioscience institute). The planning process was based on making a large number of unsubstantiated assumptions about stakeholder requirements and conditions of the envisioned funders for making funding commitments. The process of going through the planning process with the DSM software tool allowed us to rigorously focus on the information dependencies instead of on logistical steps. The process proved effective at building the necessary assumption verification tasks into the

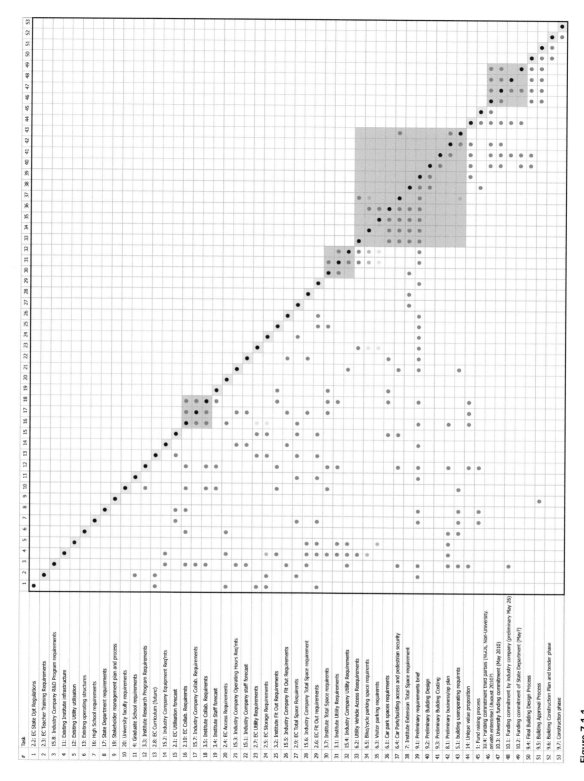

Figure 7.1.1
DSM with four coupled blocks after initial partitioning.

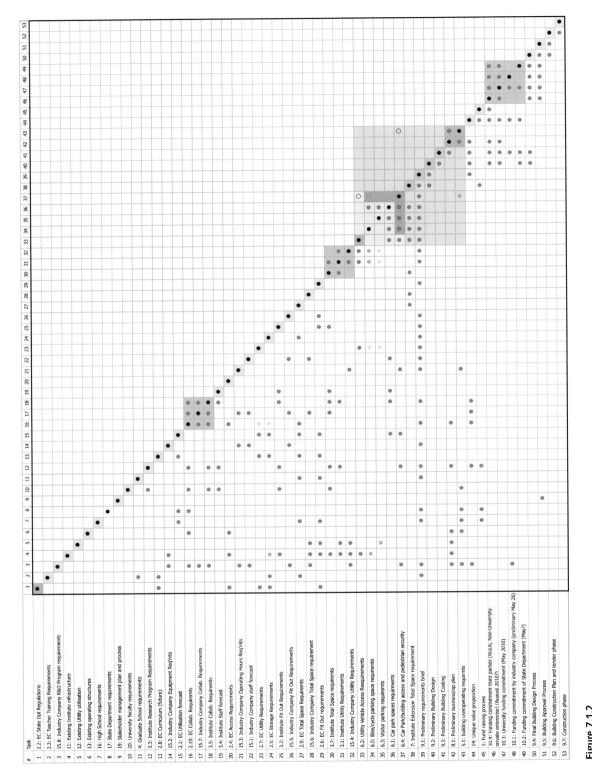

Figure 7.1.2
Simplification of the third coupled block was achieved by tearing two dependencies.

project, and it resulted in a project task sequence that could be implemented by the project team.

This process highlighted unexpected issues relating to parking, pedestrian access, private and utility vehicle access, and issues relating to collaboration between university and commercial scientists, school students, lecturers, and teachers within the facility.

During verification of the various stakeholder requirements and assumptions that had been made previously, it became evident early in the development phase that the requirements of the commercial company had changed substantially during the previous six months, meaning that they no longer needed additional facilities or space and, therefore, decided not to go ahead with co-funding the building extension. Due to the interdependencies between the commitments of the four envisioned funders (the last coupled block), this led to abandonment of the original plan after execution of the first three tasks of the project. The project was then replanned into a much smaller project to accommodate only the science high school.

Reference

Yassine, A., D. Falkenburg, and K. Chelst. 1999, September. Engineering Design Management: An Information Structure Approach. *International Journal of Production* 37 (13):2957–2975.

Example 7.2   Intel Microprocessor Product Development

Contributors

Steven Eppinger and Sean Osborne
Massachusetts Institute of Technology

Problem Statement

Intel Corporation, one of the world's leading semiconductor companies, introduced the first general purpose microprocessor to the market in 1971. By the 1990s, Intel had become the largest manufacturer of microprocessors, along with a range of other computer chips. One of the microprocessor chips from Intel is shown in figure 7.2.1 (much more recent than the ones studied in this example). For this DSM application, we were asked to help Intel to do the following:

1. Better understand their complex process for microprocessor development in general,
2. Reduce the microprocessor product development lead time, and
3. Reduce the unpredictability experienced in the microprocessor product development lead time.

Data Collection

Sean Osborne, a master's student in MIT's Leaders for Manufacturing program, was assigned to a six-month internship at Intel in 1992, with the above goals. Over a four-week period, he met with approximately 25 experienced engineers and managers to learn Intel's

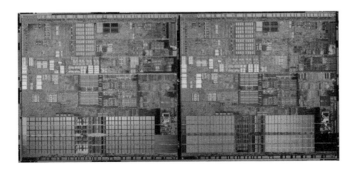

Figure 7.2.1
A dual-core microprocessor chip from Intel (courtesy of Intel Corp.).

microprocessor product development process. Based on these interviews, he represented the process using a process architecture DSM model.

Model

The DSM model shown in figure 7.2.2 represents Intel's existing product development process in 1992, as explained by the managers interviewed. The model contains 60 activities, listed on the left side of the DSM. Green shaded marks below the diagonal represent flows of information from earlier activities to later ones. Yellow shaded boxes along the diagonal and pink shaded feedback marks above the diagonal represent two types of iterations—planned and unplanned—as discussed below. This model does not show any optimizations, suggestions, or improvements made as a result of this DSM application.

Results

This was one of the early industrial applications of DSM and helped to illustrate the power of the process architecture DSM not only to capture a real product development process in detail but also to help discover opportunities to improve the process. One of the key insights resulting from this model is to see the difference between the planned and unplanned iterations.

Planned iterations shown in the DSM (the yellow shaded boxes along the diagonal) are the places where the process requires work across several related activities such that rework is necessary in order to "get it right the first time." For example, the large block of 10 circuit design activities beginning with Functional Modeling (activity 17) involves a number of interconnected design-analysis iterations. This block is followed by another overlapping block for layout iterations and then a third block of validation iterations.

Unplanned iterations are shown in the DSM as marks above the diagonal, representing ways in which the planned PD process can fail, resulting in unplanned rework of earlier activities. (Note the applicability of failure modes and effects analysis [FMEA] to processes as well as products.) Perhaps it is not surprising that several of the unplanned iterations emerge from testing and validation activities later in the process. For example, if Thermal Testing (activity 54) determines that the chip will fail thermally, engineers then need to follow one of four iteration paths. Depending on the specific type of thermal failure identified, they would either (1) rework the manufacturing process (activity 52), (2) repeat the debugging activity (activity 35), (3) repeat the chip packaging (activity 29), or (4) redesign the functional model (activity 17). Any of these paths could potentially then include rework of several additional activities, adding up to significant project delays. In looking at this and eight similar projects at Intel, we found that 13% to 70% of process duration, with a mean of 30%, was attributable to iteration and rework.

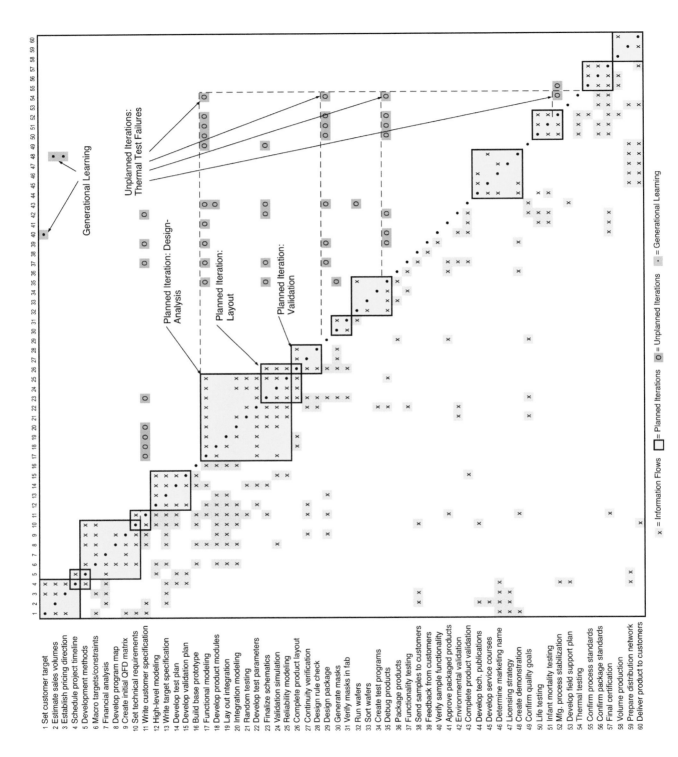

Figure 7.2.2
DSM model representing the product development process at Intel.

1 Set customer target
2 Estimate sales volumes
3 Establish pricing direction
4 Schedule project timeline
5 Development methods
6 Macro targets/constraints
7 Financial analysis
8 Develop program map
9 Create initial QFD matrix
10 Set technical requirements
11 Write customer specification
12 High-level modeling
13 Write target specification
14 Develop test plan
15 Develop validation plan
16 Build base prototype
17 Functional modeling
18 Develop product modules
19 Lay out integration
20 Integration modeling
21 Random testing
22 Develop test parameters
23 Finalize schematics
24 Validation simulation
25 Reliability modeling
26 Complete product layout
27 Continuity verification
28 Design rule check
29 Design package
30 Generate masks
31 Verify masks in fab
32 Run wafers
33 Sort wafers
34 Create test programs
35 Debug products
36 Package products
37 Functionality testing
38 Send samples to customers
39 Feedback from customers
40 Verify sample functionality
41 Approve packaged products
42 Environmental validation
43 Complete product validation
44 Develop tech. publications
45 Develop service courses
46 Determine marketing name
47 Licensing strategy
48 Create demonstration
49 Confirm quality goals
50 Life testing
51 Infant mortality testing
52 Mfg. process stabilization
53 Develop field support plan
54 Thermal testing
55 Confirm process standards
56 Confirm package standards
57 Final certification
58 Volume production
59 Prepare distribution network
60 Deliver product to customers

Generational Learning

Unplanned Iterations:
Thermal Test Failures

Planned Iteration: Design-Analysis

Planned Iteration: Layout

Planned Iteration: Validation

x = Information Flows   □ = Planned Iterations   o = Unplanned Iterations   • = Generational Learning

In addition to the planned and unplanned iterations, the DSM model also shows three instances of generational learning. This is another type of iteration in which something is discovered during the PD process, but it is too late to incorporate the necessary changes into the current product. In these cases, the information is passed along to the next generation of the product, where it may be used in a timely manner.

As a direct result of this DSM analysis, Intel understood the tremendous impact of the unplanned iterations on the variance in project schedules. They then embarked on an effort to address each of the unplanned iterations revealed in the DSM model. Some of the process changes involved resequencing activities, adding new process steps, and allocating different resources to certain activities.

References

Eppinger, Steven D. 2001, January. Innovation at the Speed of Information. *Harvard Business Review* 79 (1):149–158.

Osborne, Sean M. 1993, June. *Product Development Cycle Time Characterization Through Modeling of Process Iteration*. Master's thesis, Massachusetts Institute of Technology, Cambridge, MA.

Example 7.3   Strategy Development Process for Meat & Livestock Australia

Contributors

Elke Scheurmann
Rapid Invention Pty Ltd.

Lewis Atkinson
Meat & Livestock Australia

Problem Statement

Meat & Livestock Australia (MLA) is an industry-owned company working in partnership with industry and government to achieve a profitable and sustainable red meat and livestock industry. MLA provides R&D and marketing services to the Australian red meat industry.

In 2009, MLA began the scoping process for a new knowledge and information management system, along with an upgrade of its existing intranet. The design and content of the new system needed to support existing MLA processes and work flows and allow MLA staff to access required information relevant to specific process steps. Because many processes had never been defined formally, new employees had to rely to a large extent on the tacit knowledge of experienced managers to become competent in their jobs. We wanted to find out whether we could use the DSM methodology to define and optimize work processes and workflows so that their related information could be built into the intranet and accessed at the appropriate step. As an example process, we used the Research Program Strategy Development (RPSD) process, which had not yet been formally defined.

Programs are the primary way that MLA structures its work and delivers results to stakeholders. The MLA Annual Operating Plan is structured across a small number of broad strategic themes within which each program is articulated. The three basic qualifications that a program needs to meet before it is funded are industry benefit, innovation, and stakeholder engagement. An MLA R&D program investment is often a co-investment with other partners and can be up to $15 to $20 million over several years.

Data Collection

We selected two novice program managers (PMs) with less than one year of job experience, who had never planned a program before and posed the following question to them: "How would you go about planning a completely new research program in your area if you were asked to do that tomorrow?" We used the ProjectDSM software (www.ProjectDSM.com) to enter all the activities the PMs would undertake without any particular order of execution. As it became apparent that the novice PM planning process did not cover the complete required RPSD (Hubbard 2008), we then used our experience in R&D and business planning to fill in the gaps by adding further activities. We then defined the information dependencies between each of the strategy planning activities. The software automatically created and sequenced a DSM and identified a number of coupled blocks. We optimized the sequence of the activities within each block by either promoting or delaying activities or by tearing dependencies.

The final optimized task sequence was discussed with the novice PMs to identify and correct any issues of implementation.

Model

To arrive at an optimized strategy development process for MLA PMs, it was necessary to make a number of assumptions about the outcomes of several tasks. Figure 7.3.1 shows the initial matrix with highlighted coupled blocks after adding all required RPSD activities.

The largest coupled block was simplified first by tearing the dependency between appropriate prioritization of research areas and the overall impact on the meat industry value and supply chains. By making this one assumption, this block was broken into two smaller ones relating to the ex-ante program evaluation steps of defining appropriate research questions, identifying research capabilities and collaborators, and defining and apportioning the commercial opportunities and benefits from the research program. We then continued to tear the smaller blocks, with a total of 17 assumptions. The final DSM is shown in figure 7.3.2.

Results

The process of using the DSM methodology for analyzing a complex work process such as strategy development proved successful for MLA. It yielded a process that could be followed even by novice PMs, provided they had access to appropriate support material on the intranet about the requirements and deliverables of each task in the process.

PMs found that defining the process and its activities was valuable. They realized that they did not have sufficient information available to them to make appropriate planning decisions and take all necessary steps to arrive at a solid program strategy without having to consistently fall back on consultation with their immediate managers.

In line with the current literature about strategy development and implementation, our DSM analysis also highlighted the interdependencies between the environmental/market analysis, capability analysis, and strategy development steps.

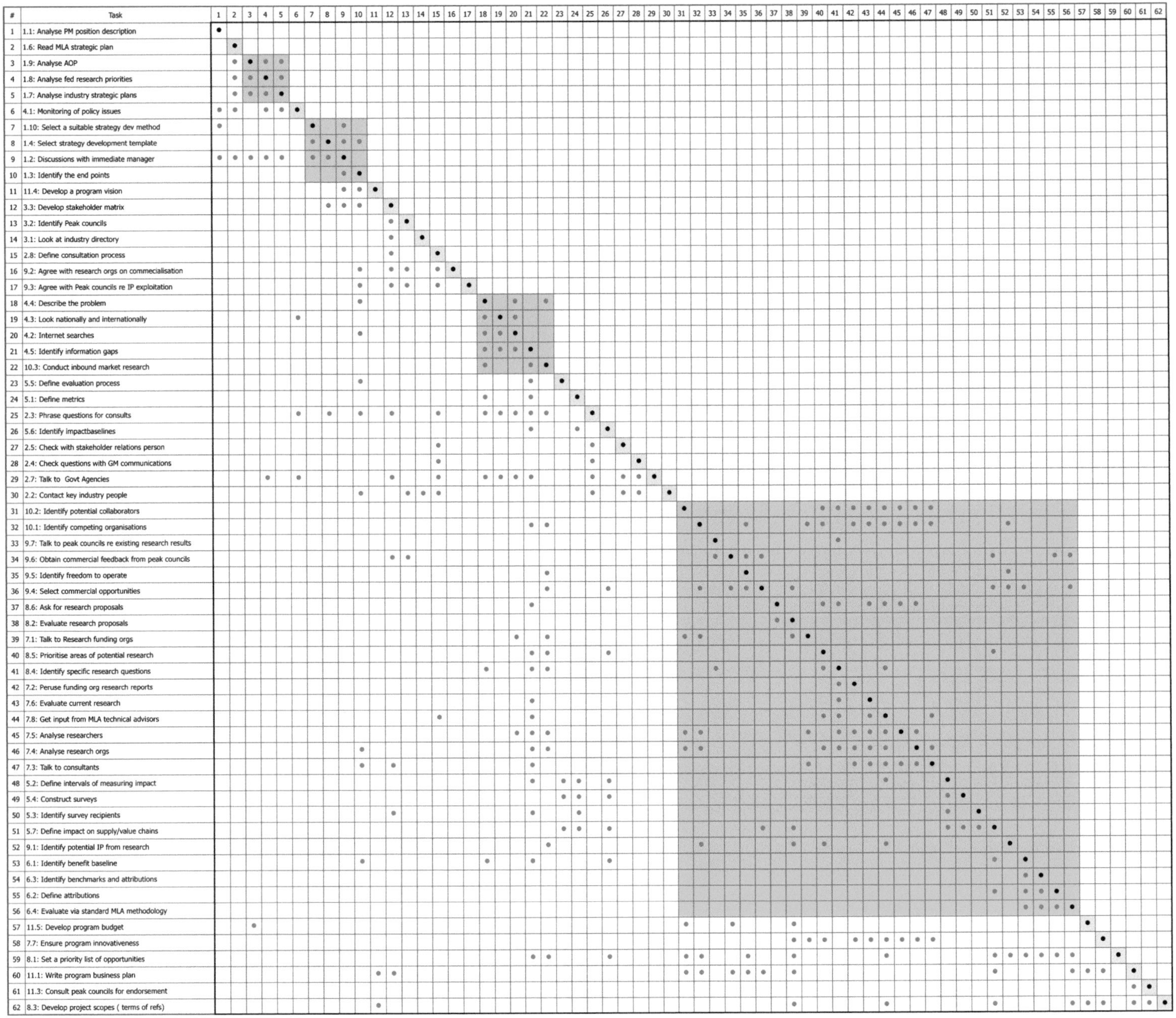

Figure 7.3.1
RPSD process DSM after initial partitioning.

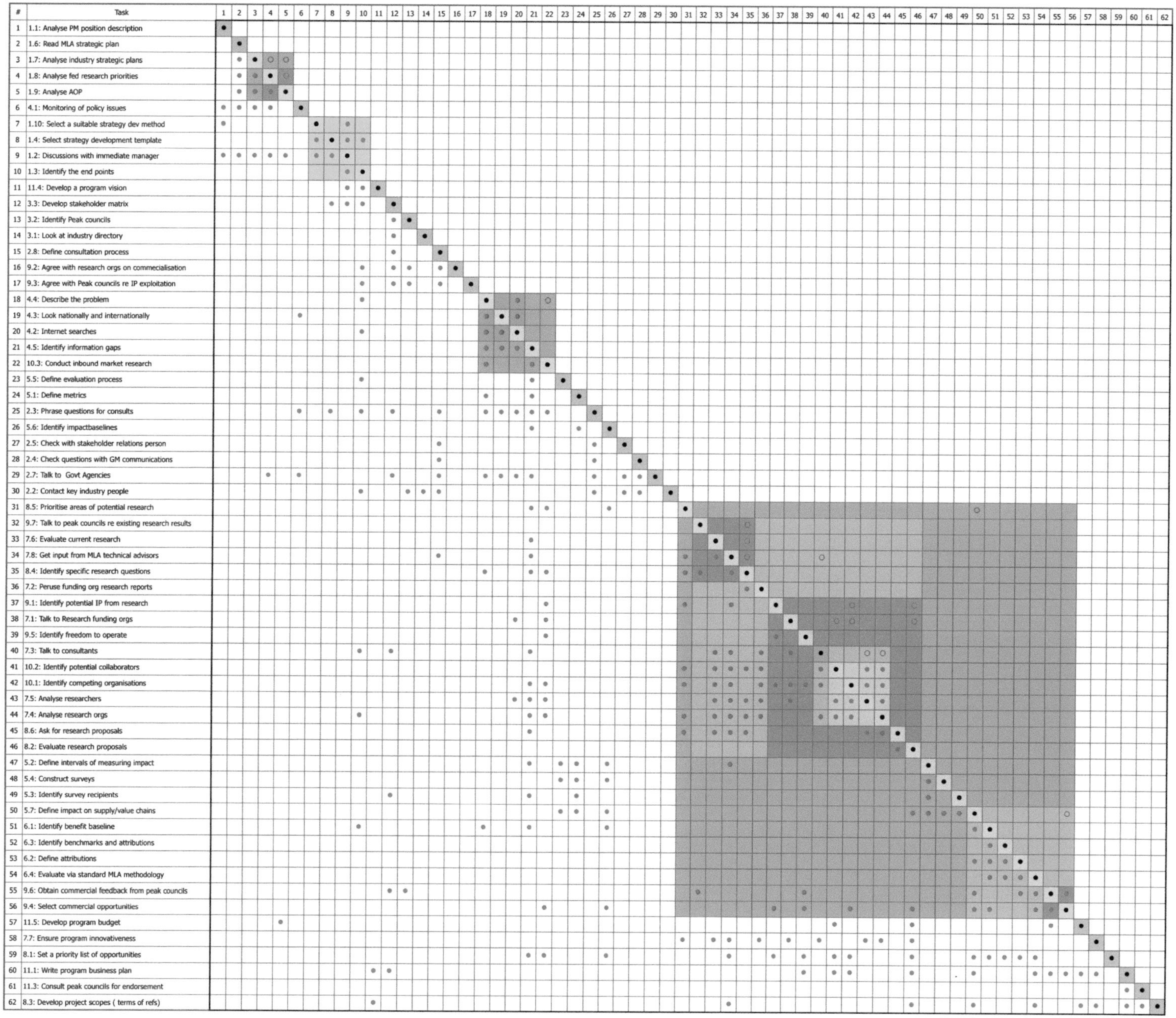

#
Task
1 1.1: Analyse PM position description
2 1.6: Read MLA strategic plan
3 1.7: Analyse industry strategic plans
4 1.8: Analyse fed research priorities
5 1.9: Analyse AOP
6 4.1: Monitoring of policy issues
7 1.10: Select a suitable strategy dev method
8 1.4: Select strategy development template
9 1.2: Discussions with immediate manager
10 1.3: Identify the end points
11 11.4: Develop a program vision
12 3.3: Develop stakeholder matrix
13 3.2: Identify Peak councils
14 3.1: Look at industry directory
15 2.8: Define consultation process
16 9.2: Agree with research orgs on commercialisation
17 9.3: Agree with Peak councils re IP exploitation
18 4.4: Describe the problem
19 4.3: Look nationally and internationally
20 4.2: Internet searches
21 4.5: Identify information gaps
22 10.3: Conduct inbound market research
23 5.5: Define evaluation process
24 5.1: Define metrics
25 2.3: Phrase questions for consults
26 5.6: Identify impactbaselines
27 2.5: Check with stakeholder relations person
28 2.4: Check questions with GM communications
29 2.7: Talk to Govt Agencies
30 2.2: Contact key industry people
31 8.5: Prioritise areas of potential research
32 9.7: Talk to peak councils re existing research results
33 7.6: Evaluate current research
34 7.8: Get input from MLA technical advisors
35 8.4: Identify specific research questions
36 7.2: Peruse funding org research reports
37 9.1: Identify potential IP from research
38 7.1: Talk to Research funding orgs
39 9.5: Identify freedom to operate
40 7.3: Talk to consultants
41 10.2: Identify potential collaborators
42 10.1: Identify competing organisations
43 7.5: Analyse researchers
44 7.4: Analyse research orgs
45 8.6: Ask for research proposals
46 8.2: Evaluate research proposals
47 5.2: Define intervals of measuring impact
48 5.4: Construct surveys
49 5.3: Identify survey recipients
50 5.7: Define impact on supply/value chains
51 6.1: Identify benefit baseline
52 6.3: Identify benchmarks and attributions
53 6.2: Define attributions
54 6.4: Evaluate via standard MLA methodology
55 9.6: Obtain commercial feedback from peak councils
56 9.4: Select commercial opportunities
57 11.5: Develop program budget
58 7.7: Ensure program innovativeness
59 8.1: Set a priority list of opportunities
60 11.1: Write program business plan
61 11.3: Consult peak councils for endorsement
62 8.3: Develop project scopes (terms of refs)

Figure 7.3.2
RPSD process DSM showing the final task sequence after tearing the large block of coupled tasks.

Noting the need to make certain key assumptions allowed MLA to define specific information and templates for PMs to help them to cope with the complex interrelationships between the tasks.

As a consequence of having used the DSM methodology on this project, MLA is now using it to simplify several other process workflows in their organizations prior to specification and implementation of a Digital Asset Management system and supporting intranet resources.

Reference

Hubbard, Graham, John Rice, and Paul Beamish. 2008. *Strategic Management: Thinking, Analysis, Action.* Frenchs Forest N.S.W.: Pearson Education Australia.

Example 7.4   Real Estate Development at Jones Lang LaSalle

Contributors

John Sullivan, Benjamin Bulloch, and David Geltner
Center for Real Estate, Massachusetts Institute of Technology

Problem Statement

Real estate development, as with any capital-intensive project, involves a complex process in which a developer looks to meet a market demand at a particular moment in time for an economically viable cost. During this process, tasks are completed and the information produced is synthesized into other related tasks. This information flow iteratively changes the development process and ultimately shapes the end product. The outcome of each task is never fully certain at the beginning. Both internal and external events in the process can result in planned or unplanned changes, making the process of development highly iterative. Through these numerous iterations, information is collected, analyzed,

Figure 7.4.1
4 Van de Graaff Drive, an office building in Burlington, Massachusetts, developed by Jones Lang LaSalle (courtesy of Jones Lang LaSalle).

and disseminated to other project participants. It is the role of the real estate developer to understand and effectively manage the information flows among the dozens of project stakeholders.

Throughout the real estate development life cycle, the project team executes many tasks as they work toward construction, completion, occupancy, and financial stabilization. Although each of these tasks varies in length, cost, and desired outcome, they can be generally grouped into five functions:

1. Market and competitive analysis (*Marketing*)

2. Physical and design analysis (*Design*)

3. Political and legal analysis (*Political*)

4. Financial analysis (*Financial*)

5. Project management

Although many tasks interact and share information within a function, many tasks share and receive information to and from tasks of different functions. The process of managing intra-functional exchanges versus managing cross-functional exchanges of information can be different. When working on two tasks within the same function, goals are more easily understood, and the tasks are generally completed by the same group or type of people. When information is shared between tasks of different functions (e.g., impact of the design on financial returns), miscommunication is more likely to occur if the exchange of information is not handled carefully. Project and financial risk is more likely in these types of interactions. Identifying where cross-functional interactions occur can help determine where additional management and oversight may be required. By highlighting these types of interactions, a developer can more efficiently utilize time and resources, reducing the risk that is inherent in any real estate development process.

Data Collection

We worked with the Boston office of Jones Lang LaSalle, a leading real estate development firm, one of whose projects is shown in figure 7.4.1. We began by exploring the range of disciplines required and tasks executed during the real estate development process. Through interviews of individual team members and the group as a whole, we identified the standard tasks involved in a typical project and which of the five primary functions owned each task. We then conducted interviews to identify the information exchanges required to execute each task and to better understand how the tasks were completed. With these data, we created a baseline DSM, shown in figure 7.4.2, to represent the general interactions and information flows for a typical real estate development project.

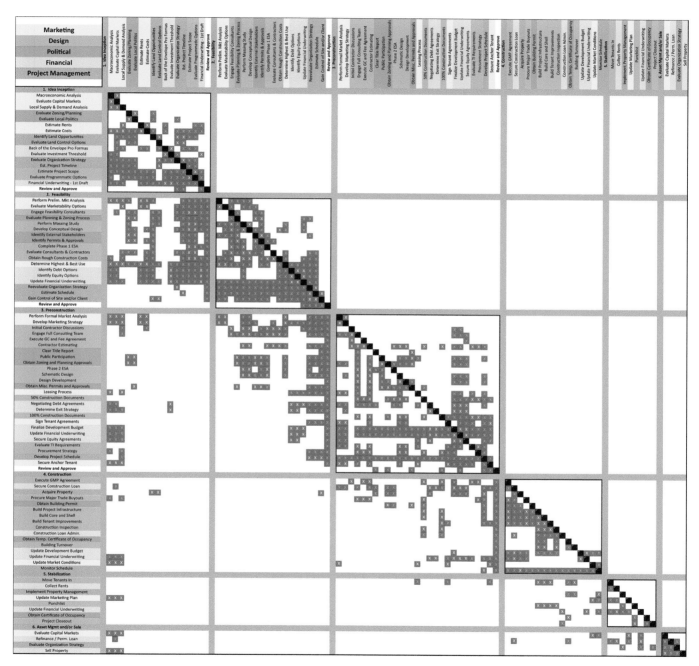

Figure 7.4.2
Functional-interaction process DSM for a real estate development project.

Model

After choosing a desired level of granularity for analysis and completing the data collection process mentioned earlier, we created a real estate development DSM consisting of 91 individual tasks and 1,148 information exchanges, which are noted by X marks in the DSM. The 91 tasks were grouped according to the six stages of the development process and also identified with one of the five functions and labeled accordingly with different colors.

We also color coded the information exchange marks in the DSM to distinguish intra- and cross-functional interactions. Green marks represent tasks that interact with tasks in the same functional group (e.g., two *Marketing* tasks or two *Political* tasks), while red marks represent tasks that interact across functions (e.g., a *Marketing* task provides information to a *Financial* task). This allows a quick and clear understanding of where and when these more challenging, cross-functional interactions may occur. Interestingly, our analysis found in this case that 69.5% of the interactions were across functions (red in the DSM).

Results

By adding color to the DSM tasks and interactions, this model adds additional depth to the matrix, yielding important insights. As explained previously, task interactions that occur within the same function (shaded in green) are likely to occur between groups that "speak the same language" or have disciplinary affinities. This information flow is more likely to be smooth with a lower risk of miscommunication or conflicting goals.

Our analysis identified the multidisciplinary interactions (shaded in red), which occur between teams that have less in common or are less familiar with each other's professional work. For example, a zoning attorney may have difficulty communicating a complicated zoning board ruling to the architecture team, which must then translate this information into design changes. This type of information flow may require greater attention by the project manager, who is responsible for successfully synthesizing the flow of information both within and across the functional boundaries. These may be more risky or more costly interactions with a higher likelihood of unnecessary iterative cycles. The architecture team in this example may draft five versions of a design until it finally matches the zoning board's requirement. Although multiple iterations may help ensure a more comprehensive process, it comes at the expense of both time and money. An efficient balance between thoroughness and timely completion must be understood and implemented by the developer to ensure optimal results. This model, which we call a *Functional-Interaction Process DSM*, provides a useful tool for managers to visualize one important source of preventable risk in large projects.

References

Bulloch, Benjamin, and John Sullivan. 2009, September. *Application of the Design Structure Matrix (DSM) to the Real Estate Development Process*. Master's thesis, Massachusetts Institute of Technology, Cambridge, MA.

Bulloch, Benjamin, and John Sullivan. 2010, July. Information—The Key to the Real Estate Development Process. *Cornell Real Estate Review* 8:78–87.

Example 7.5   Biogen Idec Drug Development

Contributor

Anshuman Tripathy
Indian Institute of Management, Bangalore

Problem Statement

Biogen Idec is a leading biopharmaceutical company involved in drug development in the areas of autoimmune disorders, neurological disorders, cancer treatment, and so on. It was formed in 2003 by the merger of Biogen and Idec. Their key drugs include Avonex (figure 7.5.1), Tysabri, and Rituxan. As a response to increased activity in their development pipeline, Biogen Idec introduced a formal drug development process in 2003. In 2004, a project team was tasked to review the new process for opportunities to smooth the transition from the end of its Research Phase to the start of its Development Phase (an in-between phase known as R-to-D Transition). The team developed a DSM to represent the drug development process flow, including its tasks, the dependencies among the tasks, and the prevalent iterations. Analysis and discussion of the DSM identified a change in the drug development process that could help improve the process during the R-to-D Transition Phase.

Data Collection

Based on a series of interviews with scientists and managers at Biogen Idec in 2005, Anshuman Tripathy (then a PhD student at MIT) worked with the Biogen Idec team to

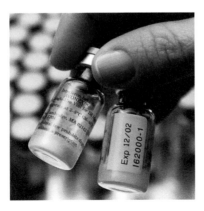

Figure 7.5.1
Avonex multiple sclerosis treatment by Biogen Idec (courtesy of Biogen Idec).

develop the DSM model for the early drug development process leading to the start of phase 1 clinical trials. This 145-task process DSM covered the process from strategic opportunity generation to the injection of the drug into the first patient in phase 1 clinical trials. The DSM process model was verified by various scientists and managers at Biogen Idec.

Model

Figure 7.5.2 shows a condensed form of the DSM model with only 53 tasks spanning three phases of the processes: Research, R-to-D Transition, and Early Development. This DSM shows two groups of activities during the Research Phase: Prospect Evaluation and Candidate Identification. Molecular antibody development, toxicology tests, and pharmacokinetics are some of the key activities that take place in this early stage of the process. This is followed by the R-to-D Transition Phase, during which the preclinical development plans are developed. Thereafter, the Early Development Phase begins with a group of activities known as the investigational new drug (IND) Enabling Track, which leads to the start of phase 1 clinical trials. Animal trials (a requirement for establishing safety standards of the drug), technology transfer, confirmation by manufacturing and quality control, and initial meetings with the Food and Drug Administration (FDA) are important activities that take place during the Early Development phase.

Results

The DSM identified several iterations within and across the phases described earlier. However, through our analysis and discussions with the team, one of these iterations provided the clearest opportunity to intervene and improve the process. The pre-IND meeting with the FDA (task 46) was part of the IND Enabling Track. The purpose of this meeting was to get formal feedback from the FDA on the IND-enabling toxicology study plan and the proposed phase 1 clinical plan and protocol. In most cases, this meeting would result in the FDA asking Biogen Idec to revise its plans, requiring the program team to return to the FDA later for approval of the plan (task 34). We recommended having the R-to-D Transition Phase gate take place only after the pre-IND meeting is completed and firm plans leading to IND-filing are drawn up. To enable this change, several other tasks (tasks 36, 40, 41, 42, 43, 44, 45) needed to be advanced from the Early Development Phase to the R-to-D Transition Phase. Three new steps also needed to be added: providing resources to CMC/Clinical/PCDS (task 54), revision of IND-enabling toxicity study protocol (task 55), and revision of CDP and phase 1 clinical protocol concept (task 56). These were required to support senior management approval for R-to-D Transition (task 34). These process changes are represented in the DSM shown in figure 7.5.3.

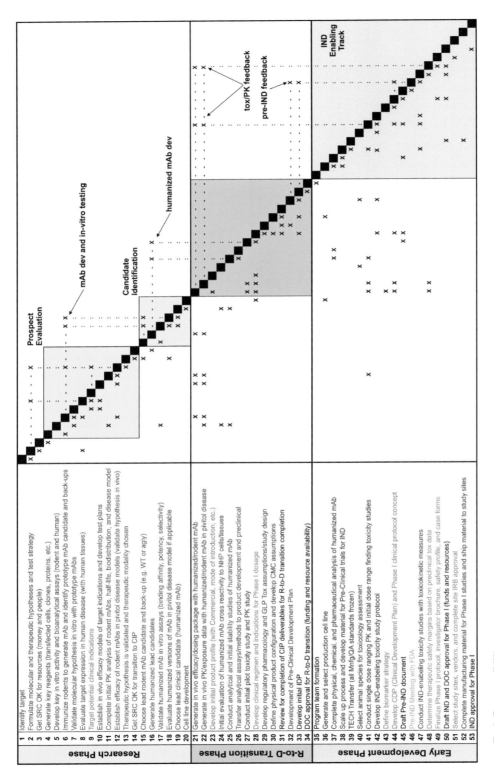

Figure 7.5.2
Existing process flow DSM for the early phases of the Biogen Idec drug development process, spanning Research, R-to-D Transition, and Early Development.

Task Responsibility

Research
Product Development
Nonclinical
Clinical
Regulatory
Transiton Team or Program Team

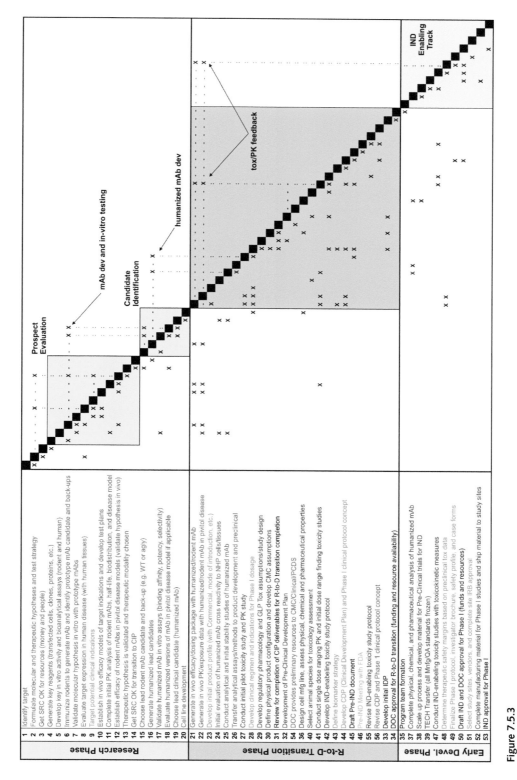

Figure 7.5.3
Proposed process flow DSM for the early phases of the Biogen Idec drug development process.

The recommended changes provided two key benefits:

1. There is a physical deliverable at the R-to-D gate. The existing process had a list of deliverables regarding information flow from Research to CMC, PCDS and Clinical, but there was no outside feedback, no particular experiment/testing, no event, and so on that verified the quality of output from Research to Development at this phase gate.

2. Senior management would now approve (task 34) firmer timing plans and resource requirements, leading to IND filing and firm plans for dosing first subject. Although the number of phase gate meetings between the program team and the senior management would remain the same, it would reduce the need for the program team to revert to senior management for approval of revised timing/resources subsequent to the pre-IND meeting (outside the phase gate review).

Reference

Tripathy, Anshuman. 2005, October. *Application of Design Structure Matrix (DSM) to Early Drug Research and Development Process*. 7th International Dependency Structure Matrix (DSM) Conference, Seattle, WA.

Example 7.6   Boeing UCAV Design Process Modeling and Simulation

Contributor

Tyson Browning
Neeley School of Business, Texas Christian University

Problem Statement

In the late 1990s, Boeing designed various unmanned combat aerial vehicles (UCAVs) for the U.S. military (figure 7.6.1). Each successful UCAV design would evolve through several phases, beginning with the Conceptual and Preliminary Design phases. Each phase involved several disciplines, each evaluating a design from their own perspective and then sharing information, making the process highly iterative. We built a process model to increase understanding of these iterations and their implications. In particular, we wanted to simulate the effects of the process architecture on the project's cost, duration, and risks.

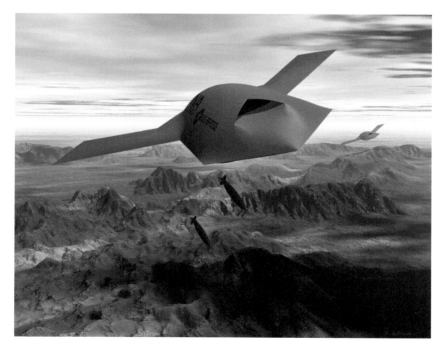

Figure 7.6.1
Artist's concept drawing of the X-45A aircraft, one of several UCAV designs developed by Boeing for the U.S. Joint Unmanned Combat Air System (courtesy of The Boeing Company).

Data Collection

In 1997, working on site at Boeing as a researcher from the Lean Aerospace Initiative at MIT, I built the model and simulation as part of my doctoral research. With input and support from Harold (Mike) Stowe and several other Boeing employees, we identified an initial set of activities comprising the Conceptual and Preliminary Design phases as well as one or more individuals with expertise in each of the activities. Over several weeks, I conducted face-to-face interviews with most of these individuals and received additional inputs via a survey form. The initial model and results were reviewed by several Boeing employees who provided additional input for verification and calibration. Full details of this process and the model are provided in the dissertation (Browning 1998).

Model

The model was built at the level of 12 Conceptual Design and 14 Preliminary Design activities, as shown in figure 7.6.2. Each phase consists of an initial activity to define design requirements and objectives (DR&O), followed by a couple of activities to create and distribute a design configuration (a design concept proposed to satisfy the DR&O). Then, several disciplines—such as aerodynamics, propulsion, stability and control (S&C), mechanical and electrical, weights, and performance—each evaluate the configuration from their own perspective. In the Conceptual Design phase, these various analyses and evaluations are pulled together in a review activity (11), which results in either a decision to proceed to the Preliminary Design phase or to iterate within the Conceptual Design phase by revising the design configuration, the DR&O, or both. The Preliminary Design phase is similar, although each discipline evaluates the proposed configuration in greater detail and some additional disciplines (such as manufacturing) are added. This phase culminates in the gathering of all data pertaining to the design for use in preparing a proposal to secure further funding for a subsequent Detailed Design phase. Note that iterations may occur within the Preliminary Design phase, but there is no formal return to the Conceptual Design phase from Preliminary Design.

The DSM includes regions showing external inputs (above) and external outputs (to the right). For this reason, this DSM adopts the IC/FBD convention discussed in chapter 6. Note that the names shown by the rows and columns of these regions identify only the name of the input or output, not its supplier or receiver (although it would be more technically correct to make them do so).

Figure 7.6.3 shows some of the additional data collected about the Preliminary Design activities. The left side of the figure shows a numerical DSM (now using the IR/FAD convention) where the off-diagonal cells indicate the *probability* of an activity's output causing any rework for the activity using it as an input. (For example, the output of activity 9 has a 20% chance of causing rework for activity 2.) The middle part of the figure

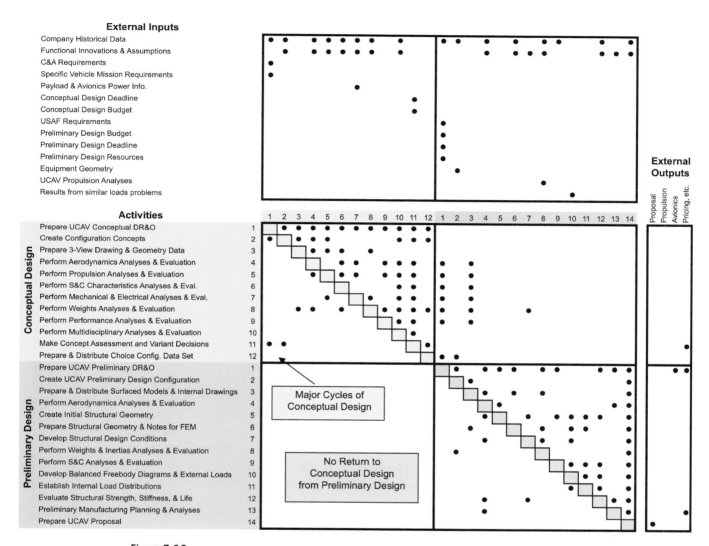

Figure 7.6.2
Process DSM model of the conceptual (yellow) and preliminary (green) design phases for UCAV development at Boeing (shown with the IC/FBD convention to facilitate the orientation of the external input and output regions).

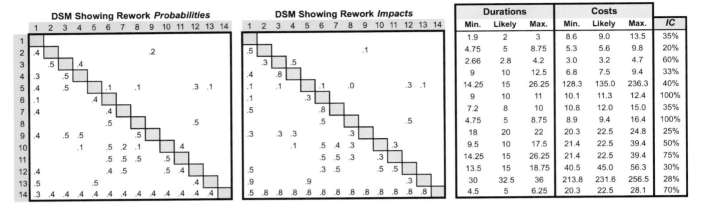

Figure 7.6.3
Rework probability and impact DSMs (IR/FAD), duration, cost, and improvement curve data for the Preliminary Design phase activities.

shows another numerical DSM where the off-diagonal cells indicate the *impact* of any such rework that occurs, where the impact is expressed in terms of the portion of the activity having to be redone. (For example, if the output from activity 9 causes rework for activity 2, then it will require activity 2 to redo 10% of its work.) In one case (activity 8 → 5), the impact of any rework was deemed negligible.

The right side of figure 7.6.3 shows the duration, cost, and improvement (or learning) curve (IC) data for each activity. The experts were asked to provide three estimates of activity duration (in work days): minimum (or optimistic), most likely, and maximum (or pessimistic). The three cost estimates (in thousands of dollars) were derived by multiplying each respective duration estimate by the resource estimates the experts also provided. Experts also supplied the improvement curve estimate, which is applied as a simple step function: For the second and any subsequent iterations of an activity, it will require $x\%$ of its original duration and cost. (For example, reworking activity 13 requires only 28% of the time and cost taken in the initial pass.) The IC helps account for common situations where design and evaluation activities build complex models, for example, but can then rerun those models much more quickly with revised inputs. Note that all data provided for public release from this project have been disguised.

Results

I built a discrete-event, Monte Carlo simulation to estimate a joint distribution of overall duration and cost for the Preliminary Design phase. The tool used each activity's sequence in the DSM (1–14) to simulate the following work policy:

- An activity must wait on any upstream activity (i.e., any activity earlier in the sequence) from which it receives direct inputs.

- An activity may proceed without any inputs it needs from downstream activities (by making assumptions about them).

We used this work policy to allow for a comparison of various activity sequences (process architectures) in the DSM.

Whenever a simulated activity finishes, the simulation checks for the possibility of rework for upstream activities. If any such rework occurs, then the simulation also checks for any additional rework caused for interim activities (called second-order rework). For example, if activity 9 causes some rework for activity 2, that change in activity 2 has a 30% chance of propagating to (causing second-order rework for) activity 3, and so on. Because rework checks are made each time an activity finishes (whether it is being worked for the first time or not), higher orders of rework, although rare, are also captured by the simulation.

Using Latin Hypercube sampling, the simulation finds a duration and then a cost (90% correlated with the duration sample) for each activity at the start of each run. These durations and costs are added as the simulation progresses until all activities (and any rework) are finished, resulting in a total duration and cost for the project. The project is simulated repeatedly in batches of 100 runs until the mean and variance of both the duration and cost distributions stabilize to within 1%. Generating stable output distributions for the UCAV Preliminary Design process typically required about 1,100 to 1,400 runs (requiring at most a few seconds on modern computers). For further details of the simulation's implementation, see Browning and Eppinger (2002).

Figure 7.6.4 shows an example Gantt chart generated from a single simulation run. This instance demonstrates several interesting characteristics of the simulation. First,

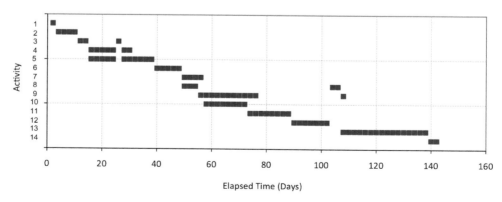

Figure 7.6.4
Example Gantt chart from simulation of the process DSM (adapted from Browning and Eppinger 2002).

note that activities 4 and 5 begin together because they do not depend on each other. Activity 4 finishes first and causes some rework for activity 3, which immediately begins again. Activity 5, which had not yet finished, halts its work because it also depends on activity 3. Activity 5 resumes once activity 3 finishes. Because activity 4 also resumes at this point, we find that the rework of activity 3 had also caused some second-order rework for activity 4. Activity 12 also caused some rework for activity 8, which in turn caused some (second-order) rework for activity 9. However, this additional work for activity 9 could proceed in parallel with activity 13, which does not depend on activity 9. Finally, note activity 13's large contribution to the critical path and the project's overall duration.

Figure 7.6.5 shows a contour plot of the joint duration-cost distribution, where the shading represents the frequency of a simulated outcome. The two straight, crossing lines represent the budget of $630k (vertical line) and the deadline of 130 days (horizontal line). Thus, project managers would ideally prefer the majority of the distribution to fall in the window on the lower left, where outcomes meet or exceed both goals. However, as it stands, 51% of the project's outcomes overrun the budget, and 67% of the outcomes miss the deadline. (These statistics are easily gathered from the cumulative form of the distribution, the integral of the distribution shown in figure 7.6.5.)

Many process improvement methodologies suggest ways to perform activities faster and cheaper. Although these can be beneficial, the DSM simulation provides insight on ways to improve the overall process *without changing the cost or duration of any individual*

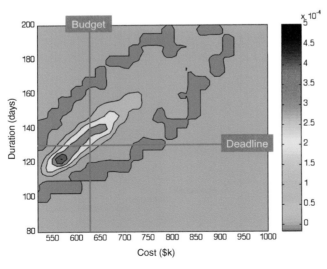

Figure 7.6.5
Joint cost and duration distribution of outcomes from the simulated process (adapted from Browning and Eppinger 2002).

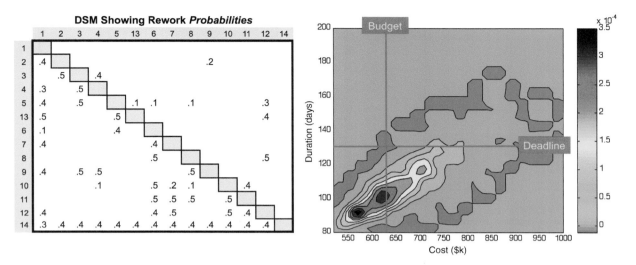

Figure 7.6.6
The improved process resulting from moving one activity upstream (adapted from Browning and Eppinger 2002).

activity. It turns out that process architecture provides a powerful lever for managers to control project time and cost. One aspect of this power, *iterative overlapping*, is first demonstrated by example. In figure 7.6.6, we move activity 13 upstream (to the sixth position), thus allowing it to begin without the input from activity 12 (making an assumption about it instead), even though this increases the likelihood of rework by adding a feedback mark in the DSM. (The impact DSM, not shown in figure 7.6.6, is also changed accordingly.) The right side of the figure shows the result of this change in the distribution of cost and duration outcomes. Although the frequency of cost overruns has increased slightly to 61%, the number of schedule overruns has been reduced substantially to 7%.

How did moving activity 13 upstream provide such a large boost in project speed? Figure 7.6.7 illustrates the effect of taking a long activity with a big improvement curve (like activity 13, which contributes a lot to project duration, as shown in figure 7.6.4, and has an IC of 28%) off the critical path and letting it get started early. Although the chances of rework increase (and along with it the cost) because of the assumption(s) being made in lieu of final information for inputs, it is much faster for the overall project to have that relatively short amount of rework on the critical path rather than the full activity. Thus, although the cost in case B is slightly greater than in case A, the duration is much less: $C_B > C_A$, whereas $D_B < D_A$.

This DSM simulation led to many extensions, including those listed in the references after the first two items.

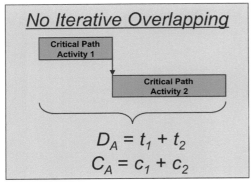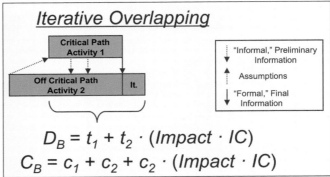

Figure 7.6.7
Iterative overlapping can accelerate a process, reducing its duration, D, while incurring only slightly greater cost, C (adapted from Browning and Eppinger 2002).

References

Browning, Tyson R. 1998. *Modeling and Analyzing Cost, Schedule, and Performance in Complex System Product Development*. PhD thesis (TMP), Massachusetts Institute of Technology, Cambridge, MA.

Browning, Tyson R., and Steven D. Eppinger. 2002. Modeling Impacts of Process Architecture on Cost and Schedule Risk in Product Development. *IEEE Transactions on Engineering Management* 49 (4):428–442.

Abdelsalam, Hisham M.E., and Han P. Bao. 2007. Re-sequencing of Design Processes with Activity Stochastic Time and Cost: An Optimization-Simulation Approach. *Journal of Mechanical Design* 129 (2):150–157.

Afsharian, Sharareh, Marco Giacomobono, and Paola Inverardi. 2008, May 13. *A Framework for Software Project Estimation Based on COSMIC, DSM and Rework Characterization*. Proceedings of the 1st International Workshop on Business Impact of Process Improvements, Leipzig, Germany, pp. 15–24.

Araki, Katsufumi. 2008, November 11–12. *Advanced Project Management Framework for Product Development*. Proceedings of the 10th International Design Structure Matrix Conference, Stockholm, Sweden, pp. 143–156.

Chen, Dong-Yu, Wan-Hua Qiu, Min Yang, and Mei-Yung Leung. 2007, September 21–25. *Activity Flow Optimization and Risk Evaluation of Complex Project*. Proceedings of the International Conference on Wireless Communications, Networking and Mobile Computing (WiCom 2007), Shanghai, China, pp. 5191–5194.

Cho, Soo-Haeng, and Steven D. Eppinger. 2005. A Simulation-Based Process Model for Managing Complex Design Projects. *IEEE Transactions on Engineering Management* 52 (3):316–328.

Gärtner, Thomas, Norbert Rohleder, and Christopher M. Schlick. 2009, October 11–13. *DeSiM—A Simulation Tool for Project and Change Management on the Basis of Design Structure Matrices*. Proceedings of the 11th International Design Structure Matrix Conference, Greenville, SC, pp. 259–270.

Huang, Enzhen, and Shi-Jie (Gary) Chen. 2006. Estimation of Project Completion Time and Factors Analysis for Concurrent Engineering Project Management: A Simulation Approach. *Concurrent Engineering: Research and Applications* 14 (4):329–341.

Jun, Hong-Bae, Hyun-Soo Ahn, and Hyo-Won Suh. 2005. On Identifying and Estimating the Cycle Time of Product Development Process. *IEEE Transactions on Engineering Management* 52 (3):336–349.

Lévárdy, Viktor, and Tyson R. Browning. 2009. An Adaptive Process Model to Support Product Development Project Management. *IEEE Transactions on Engineering Management* 56 (4):600–620.

Wynn, David C., and P. John Clarkson. 2009, August 24–27. *Design Project Planning, Monitoring and Re-Planning through Process Simulation*. Proceedings of the International Conference on Engineering Design (ICED), Stanford, CA.

Example 7.7   Skanska Hospital Development Process

Contributors

John Steele and Paul Waskett
Adept Management Ltd.

Problem Statement

Skanska, one of the world's leading design and construction companies, works in Sweden, the United Kingdom, and the United States, primarily in the building sector. As early as 1998, the Egan report to the UK construction industry highlighted the need for improved integration of the team to ensure effective management of the design process and timely and efficient delivery of construction. Like many industries, the construction sector has long struggled with integrating the independent working processes of multiple project participants. Skanska sought methods for improving their delivery of design and procurement as an integrated process and turned to Adept Management for guidance and support. Adept Management, a UK-based design management consultancy, developed the ADePT methodology (see figure 7.7.1) and software in the late 1990s to enable integration of the design and procurement process. ADePT has DSM at its heart. Here we describe its

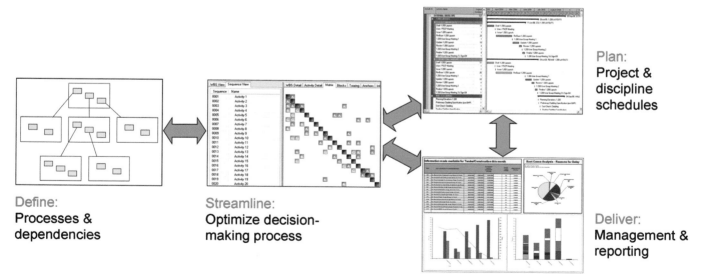

Figure 7.7.1
Process architecture DSM is at the core of the ADePT methodology.

application on a major UK hospital development by Skanska. The aim of our work was to generate an information-driven design and procurement schedule that reflected the interdisciplinary nature of the design process and could be used to optimize the overall project delivery schedule through effective alignment of the preconstruction and construction phases.

Data Collection

Adept Management consultants held an interdisciplinary design workshop for all project participants. The session was facilitated to create interaction and openness between attendees prior to a work breakdown structure (WBS) being formulated. The WBS was initially developed top-down using a discipline-based decomposition, allowing a subsystem decomposition to then follow. The resulting WBS provided the framework within which design activities could be identified and stored.

Adept Management consultants then held meetings with each of the design disciplines independently to populate the deepest level of the WBS with design activities. These were facilitated using a data library of design activities (and associated information dependencies) that had been developed and validated through research and multiple industrial applications. Where activities did not exist in the data library, they were captured and embedded within the data set. All activities were determined using this approach, and a subsequent review of the embedded information dependencies by each design discipline resulted in a validated and agreed process model. These data were then analyzed using the DSM functionality of the ADePT software.

Model

The DSM shown partially in figure 7.7.2 comprises more than 2,500 design and procurement activities at the deepest level of the WBS. The ADePT software uses a stepwise approach to build layers of data within the process model. This ensures that the model is populated with sufficient data to enable optimization and tearing within the DSM, as well as the additional data required (such as responsibility, duration, and effort) to enable automatic generation of the sequence in the client's scheduling tool of choice (Primavera P6 here). The ADePT software utilizes a numerical DSM with the dependencies between activities, in this case information flows, being rated on a 3-point scale based on the criticality of the information to completing the activity (as defined by the owner of each activity), with A being considered critical, B important, and C nice to have (the inference being that C-rated information can be easily assumed with little risk to the project). This enables optimization of the activity sequence based on the availability of outputs associated with the most critical dependencies.

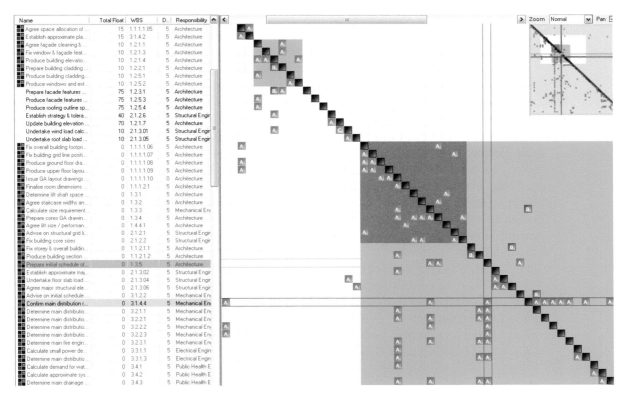

Figure 7.7.2
A block of activities in the DSM indicating interdisciplinary collaboration, displayed using the ADePT software.

The activities of the various disciplines fall into natural blocks where feedback loops exist, as seen in the DSM model. These blocks represent interdisciplinary systems that require the input of multiple perspectives to ensure coordination; in effect, the weighted optimization identifies positive periods of iteration in the design process and positions them relative to the remainder of the process.

The blocks vary in size from two activities (couples) to many hundreds of activities (large-scale interdependency) depending on the number and direction of information dependencies present in the model and the weighting that is allocated. The blocks were reviewed by the team, and those that were deemed to represent a clearly definable coordination "hot-spot" were scheduled as such within the Primavera P6 application, to be resolved either by interdisciplinary workshops or, where resource availability did not allow co-location, as stepwise processes with agreed review and rework periods. Those blocks that were not easily defined (because they contained too many activities or were, in effect, a collection of interdependent coordination events) were partitioned using an embedded sequencing algorithm. This ranks discrete dependencies between activities within each iterative loop in terms of the potential reduction in the size of the block if the dependency were removed. This then allowed the design team to interrogate the model and make collective judgments about where decisions could be made with certainty (which involved capturing the decision, assigning an owner, changing the A or B classification to a C classification, and then re-running the sequence optimization). This shared decision making helps achieve the team integration sought by all projects—which was a core objective of the Egan report recommendation.

Once the project activities were developed into a workable design and procurement process, the data were exported into the Primavera software to create the schedule. Thereafter, any changes in either the ADePT file or the P6 file would be synchronized to ensure consistency of data. This was critical in enabling change analysis during the delivery period. The optimum design sequence was linked with the construction schedule to check the interface between the design and construction process and enable the alignment of all phases of the project through an agreed procurement strategy (Waskett et al. 2010). This provided the basis for the implementation of production control principles to manage, monitor, and control the rate of information production of the integrated team. Putting the plan into action is so vital, yet it is where many projects fail, often due to the inadequacy of the definition and optimization of the design process.

Results

Skanska was familiar with DSM (as a component of ADePT) prior to this project implementation, having used the method regularly to plan the preconstruction phase of projects. However, Skanska's design teams differ in composition from project to project (the nature of the project dictating the mix of specialists required), and, as such, the modeling and review process was new to this group. The DSM-based planning process has proved to be a powerful vehicle for team engagement and synchronization in every project. In this case, the design team remained committed beyond the modeling and optimization stage to maintaining the rate of production that they defined within the integrated process—the resulting schedule being far more acceptable and reliable than an imposed timeline because it was defined from their agreed scope of work (activities), accurate information requirements, and their own terminology. The robustness of the workflow defined within the schedule and the definition of information dependencies between activities enabled sophisticated production control principles to be implemented. Consequently, weekly work plans were produced, and progress was reported using a range of performance metrics (including percentage plan complete, design days complete, and work in progress) that are rarely implemented during the preconstruction phase.

Look-ahead planning was also applied, enabling analysis of constraints, risk mitigation, and short-term process redefinition to be undertaken to adjust the process to maintain progress in line with the master project schedule. Skanska's design director on the project was in little doubt about the value of the approach, stating in a feature in *Building Magazine* (2008):

As a management operation we get to clearly see how the design team is performing against our integrated project schedule and what issues are preventing them from delivering. We also get to see trends in performance over time, which can be very informative. The technique is powerful in improving the designers' ability to deliver to the schedule and in their performance in general. It has also contributed to Skanska's efforts to continuously improve. We find ourselves in a much stronger position to deliver key procurement and construction information.

Without the DSM analysis as a component of the wider ADePT methodology, this type of dynamic process management, monitoring, and control would have been impossible to achieve, particularly given the iterative nature of the design process and the industry's propensity to sequence work based on an assumed linearity in the process—due in main to the prevalence of the critical path method.

References

Wheal, Kate. 2008, May 30. A Healthy Option. *Building Magazine* (21).

Construction Task Force. 1998. *Rethinking Construction.* London: Department of Trade and Industry, HMSO.

Waskett, P., A. Newton, J. Steele, M. Cahill, and J. Beaumont. 2010, October 20–22. *Achieving Reliable Delivery of Design Information for Procurement and Construction.* Proceedings of the 3rd International World of Construction Project Management conference, Coventry, England.

Example 7.8   Dover Motion Precision Systems Development Process

Contributors

Anshuman Tripathy
Indian Institute of Management, Bangalore

Steven Eppinger
Massachusetts Institute of Technology

Problem Statement

Dover Motion, a business unit of Danaher Corporation, produces air-bearing-based precision motion machinery (see figure 7.8.1) that is utilized in a wide range of high-tech manufacturing industries, including data storage, flat panel display, semiconductor lithography and wafer inspection, circuit board assembly, high-precision assembly, and metrology. Due to its ability to develop customized solutions based on its core air-bearing

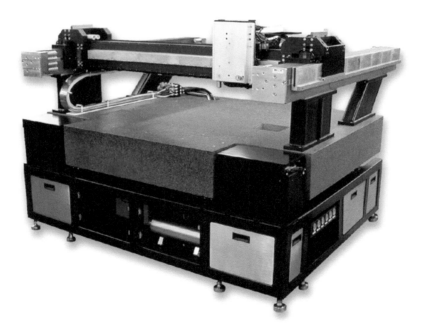

Figure 7.8.1
Precision inspection tool using Dover's air bearing technology and high-performance motion control system (courtesy of Dover).

technology, Dover has a loyal customer base that values the quality, speed, and agility with which their needs are addressed. This process architecture DSM application investigated the product development process at Dover to identify opportunities for the offshoring of development tasks to seek lower engineering labor rates.

Data Collection

An initial briefing of Dover's nascent global product development (GPD) effort was followed by interviews with Dover managers and system engineers, conducted by Anshuman Tripathy (then a PhD student at MIT) in 2006. We identified the key product development process steps followed by Dover to develop their custom precision motion systems. We then identified the sequence of process steps (tasks), the key information dependencies among these tasks, and the stage review points.

Model

The DSM in figure 7.8.2 shows the product development process architecture, beginning with the customer request and ending with shipment of the product to the customer. The six process stages and "toll gate" review positions are also identified in the DSM. The DSM marks in each row reflect the information needed by the task from prior tasks. The shaded blocks of tasks along the diagonal in the DSM represent interdependencies where tasks are performed simultaneously with mutual sharing of information. These iterative groups of tasks occur during the development of each of the key subsystems: structure (containing the core air-bearing technology), controls (electrical, hydraulic, and pneumatic systems), and software (both standard portions and elements unique to each product).

Results

Dover's initial offshoring efforts (to an engineering service provider in India) were limited to certain standard engineering tasks, such as CAD drawing and detailing. These tasks are identified in the DSM by "Outsource" in the column next to the task names and with a light shaded block in the diagonal cell. As can be seen in the DSM, these offshore tasks were coupled to several other tasks that were kept in house at Dover. The necessary iterations with the engineering service provider in India made this global process difficult to manage. As a result, the offshore, outsourced engineering service firm found the quick engineering turnaround requirements of Dover's business challenging, and Dover decided to pause the relationship.

Thereafter, Dover joined its parent Danaher Motion's offshoring efforts at their Global Development Center (GDC) in India. They started with the same content as their initial

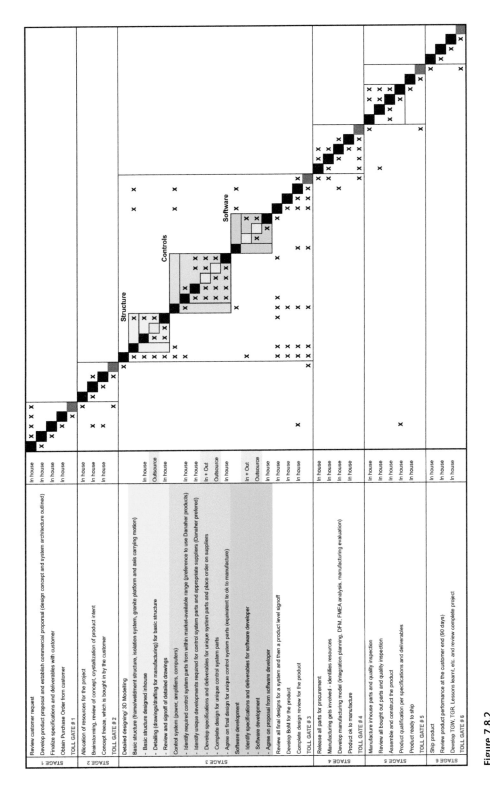

Figure 7.8.2
Process DSM model for precision equipment development process at Dover.

offshoring attempt. However, in this second attempt, project engineers at the GDC were trained on Dover's products and documented processes. This allowed for a much smoother GPD process and further development of offshore capability and responsibility.

References

Eppinger, Steven D., and Anil R. Chitkara. (2006, Summer). The New Practice of Global Product Development. *MIT Sloan Management Review* 47 (4):22–30.

Tripathy, Anshuman. 2010, January. *Work Distribution in Global Product Development Organizations*. PhD thesis, Massachusetts Institute of Technology, Cambridge, MA.

Tripathy, Anshuman, and Steven D. Eppinger. 2011, August. Organizing Global Product Development for Complex Engineered Systems. *IEEE Transactions on Engineering Management* 58 (3):510–529.

Example 7.9   Electronic Devices Product Development at Yanmar

Contributors

Minoru Okubo
Electronics Development Center, Yanmar Co., Ltd.

Yukari Arai
iTiD Consulting

Problem Statement

Yanmar Co., Ltd. is a leading Japanese manufacturer of industrial machinery, including engines, agricultural machines, construction machines, and small boats. Yanmar was founded in 1912, and its products are known to be highly reliable. In 1968, Yanmar received the Deming award for the first time in the engine industry. The Electronics Development Center (ELC) is Yanmar's dedicated R&D center responsible for the development of all electronic devices. In 2005, the ELC faced many challenges in product development (PD), such as new safety and environmental regulations, and severe global competition. It also had problems in collaboration with other divisions at Yanmar, so the success of each PD project depended heavily on the skills and experience of its project manager and members. A major initiative titled Young Energy Leads to Liberty (YELL project) was undertaken to improve Yanmar's product quality, minimize development time, and make its organization sustainable in the market. In the assessment phase of this project, a process DSM was used at Yanmar with the following objectives:

1. Visualize the dependencies among PD tasks and the actual status of rework.

2. Analyze and compare the PD process for different devices.

3. Determine the main organizational problems.

   Using DSM, Yanmar identified several PD process improvement actions and successfully achieved goals for product quality and organizational characteristics.

Data Collection

In the YELL project, iTiD Consulting supported Yanmar engineers to analyze their PD process with a DSM. Over a period of one and half months, we held about 15 two- to three-hour workshops to visualize the dependency levels among PD tasks and actual rework. We also used DSM to further analyze major unexpected rework, which led to finding the underlying problems in the Yanmar PD process.

Model

The models shown here represent the as-is PD process in 2005. We list PD tasks for vehicle electronic devices for three different modules in chronological order in figure 7.9.1. PD tasks for the same modules in figure 7.9.1 are listed separately by module in chronological order in figure 7.9.2. Schematic DSMs for four electronic devices at ELC are shown in figure 7.9.3 to compare the characteristics of each PD process.

Results

Several engineers noted in the interviews that collaboration with other engineers is not active during the PD phase at the ELC, and this can be inferred from figures 7.9.1 and

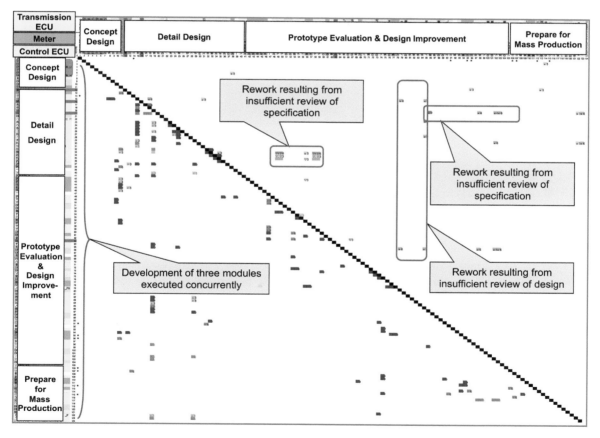

Figure 7.9.1
Process DSM for vehicle electronics (three modules developed simultaneously).

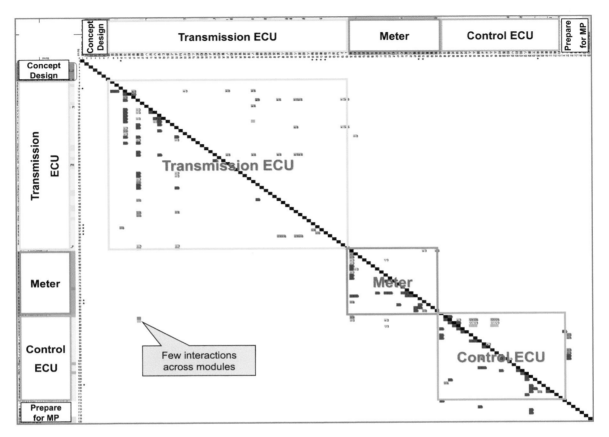

Figure 7.9.2
Process DSM for vehicle electronics (three modules developed separately).

7.9.2. Only a small number of tasks are positioned near the diagonal in figure 7.9.1, and major rework occurred in the later PD phase as a result of insufficient analysis of specifications. According to figure 7.9.2, its process seems to be a typical modular PD process with few dependency relations among modules. This implies that its PD process and organization were immature to integrate across modules. At this point, the same issues are likely to exist in the PD processes for other devices.

In the organization, the ELC functioned like a supplier of different electronic devices, and their PD targets were not well aligned with the organization as a whole. The ELC's management wanted to clearly understand the essential problems before determining strategy and action plans for redesigning the process. We also created and used DSMs to analyze the PD processes of three other devices to find similarity and uniqueness across the PD processes at the ELC. Their schematic DSMs are shown in figure 7.9.3.

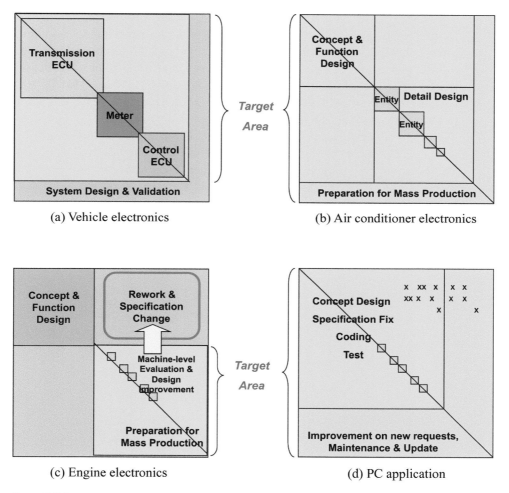

(a) Vehicle electronics

(b) Air conditioner electronics

(c) Engine electronics

(d) PC application

Figure 7.9.3
Schematic DSM models for four types of electronics products.

In figure 7.9.3a, an electronic device on a vehicle and another on an air conditioner (figure 7.9.3b) were developed mainly by combining ready-made modules, so their PD processes were similar in having a comparatively small number of iterations. However, the PD process for the air conditioner was superior because more of the iterations take place in the earlier concept design phase, as depicted in the DSM. It is believed that because more resources were spent in the early development phase, this led to a reduction in the number of coordination tasks across modules.

In figure 7.9.3c, an electronic device on an engine and a personal computer application (figure 7.9.3d) were developed mainly with new technologies, so their PD processes were similar in having a large number of iterations. With a further analysis of the DSM, we found that major rework from the later phase back to the earlier phase was the result of insufficient consideration of the product requirements before the concept design phase. For the electronic device on the engine (figure 7.9.3c), additional resources were required to resolve many issues left open from the machine-level evaluation phase.

At Yanmar ELC, DSM was used for many PD projects to visualize the dependencies among PD tasks and the sources of rework. With detailed analyses, Yanmar was able to:

1. Determine characteristics and issues in the PD processes of different devices.

2. Understand the power of DSM as a reviewing tool for their PD projects.

3. Accelerate technology innovation projects such as all-purpose ECU for different vehicles.

Reference

http://www.itid.co.jp/projects/case/006.html (in Japanese)

Example 7.10   Change Impact Analysis of Aero-Acoustic Noise Effects

Contributors

David Wynn, Nicholas Caldwell, and John Clarkson
Engineering Design Centre, University of Cambridge

Problem Statement

There is a continuing desire to reduce the environmental noise contribution from commercial jet aircraft. This involves the use of ever more sophisticated design and simulation tools to predict the effects of design changes and installation details such as flaps and engine pylons on noise levels transmitted from the aircraft to the ground. In practice, this requires design approaches that facilitate optimization at an acceptable cost. To this end, a process description for investigating jet noise aero-acoustic installation effects has been developed, and its performance in terms of cost and benefits has been investigated.

Data Collection

The project owner identified the key analysis tools required to investigate a range of performance criteria related to jet noise aero-acoustic installation effects. In addition, they identified the resources required to undertake the analysis and key decisions to be made. These resources and decisions are split across different companies involved in the process. The authors compiled this information into a process model and added key validation activities and feedback paths. A DSM and a network diagram were generated automatically from the process model by the Engineering Design Center's Cambridge Advanced Modeler (CAM) software tool.

Model

The model, illustrated in figure 7.10.1, comprises a set of design and analysis activities (represented by arrows), including key parameters and services (represented by ellipses) executed by identifiable resources (represented by people) from a variety of organizations. These activities interact via a moderate number of feed-forward and feedback links. This structure is most evident in the process flow model or equivalent force-directed network diagram and results in the blue-shaded blocks shown along the diagonal in the process DSM in figure 7.10.2. This is an unusual process DSM in that it combines activities (the block arrows from figure 7.10.1), deliverables (including design parameters and models of parts, the ellipses from the same figure), and resources (people), each as an individual row and column. However, it is interesting to observe on occasion how linking

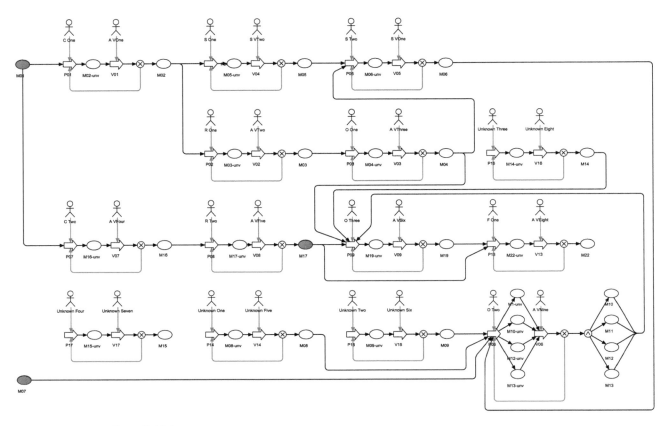

Figure 7.10.1
Design and analysis activities (arrows), utilizing and generating key parameters and services (ellipses), are executed by identifiable resources (people) from a variety of organizations.

parameters are assigned to blocks in the larger model. This model was subsequently used as the basis of a simulation, which estimates the impact of a change initiated in one or more design parameters, accounting for iterations and resource dependencies.

Results

The level of rework required to implement a change is inextricably linked to the dependencies between activities, as well as their iterations and timings. Simulation of such processes can enable the impact of a proposed change to be appreciated before committing to executing the process. The jet noise aero-acoustic installation effects model used to perform the evaluation directly reflects an existing process map created by designers. It contains iteration at two levels: (1) within each design and evaluation activity and (2)

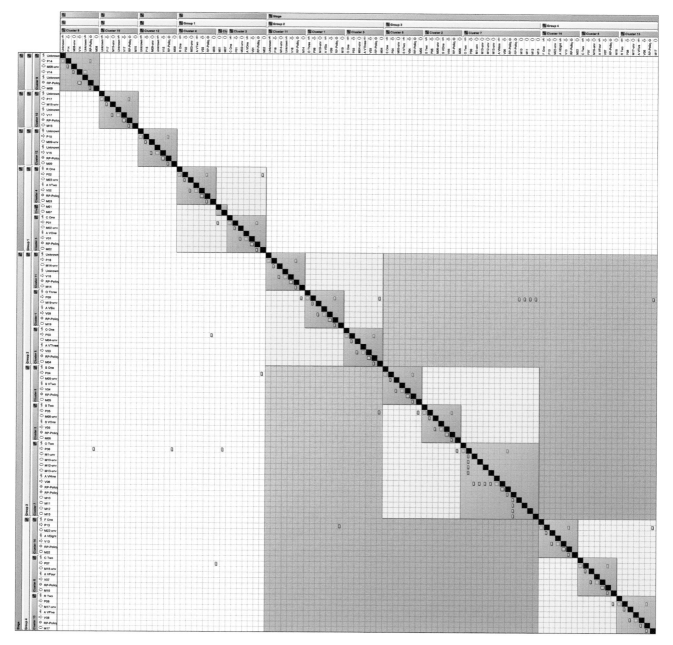

Figure 7.10.2
DSM showing coupled groups of design and analysis activities, key parameters and services, and resources.

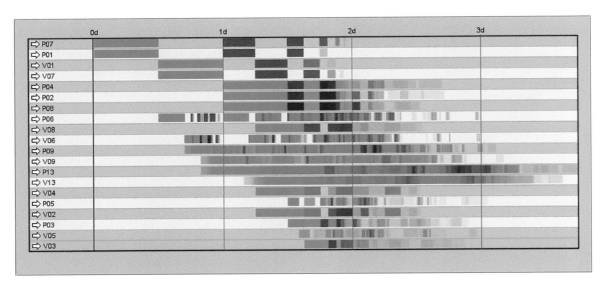

Figure 7.10.3
Design (Pxx) and analysis (Vxx) activities as elements in a probabilistic Gantt chart.

between such activities. Simulation of this process enabled the impact of the higher level iteration to be distinguished from that of the lower level activities, providing the designers with a clearer view as to the sensitivity of each activity to given changes and helping them understand the likely impact of their actions on other process stakeholders. After selecting the parameter(s) that initiate a given change, on the process map or DSM, the design rework predicted by the simulation is presented as a probabilistic Gantt chart (figure 7.10.3), which clearly identifies the design (Pxx) and analysis (Vxx) tasks and associated stakeholders that may be affected by the change and when.

References

Jarratt, Tim, Claudia Eckert, Nicholas Caldwell, and John Clarkson. 2011. Engineering Change: An Overview and Perspective on the Literature. *Research in Engineering Design* 22 (2):103–124.

Wynn, David, Seena Nair, and John Clarkson. 2009, August 24–27. *The P3 Platform: An Approach and Software System for Developing Diagrammatic Model-Based Methods in Design Research*. Proceedings of the 17th International Conference on Engineering Design, Stanford, CA.

Wynn, David, Nicholas Caldwell, and John Clarkson. 2010, May 17–20. *Can Change Prediction Help Prioritize Redesign Work in Future Engineering Systems?* Proceedings of the 11th International Design conference, Dubrovnik, Croatia.

Example 7.11   Lockheed Martin F-16 Avionics Upgrade Process

Contributor

Tyson Browning
Neeley School of Business, Texas Christian University

Problem Statement

The Lockheed Martin F-16 Fighting Falcon is a multirole jet fighter aircraft originally developed by General Dynamics for the U.S. Air Force (USAF). Since the approval of production in 1976, more than 4,400 F-16s have been produced and are in use by about 25 countries, making it the largest Western jet fighter program. Because of its ongoing popularity, Lockheed Martin continues to develop upgraded systems for the aircraft. A view of the F-16 cockpit is shown in figure 7.11.1.

In 2000, Lockheed Martin sought a way to isolate and manage the additional process iterations and the consequential increases in cost and lead time caused by delays in customer-furnished equipment (CFE), an external input provided by the USAF to the upgrade process. For example, upgrading the F-16 weapons launch capabilities requires that any new weapons be available for integration and testing. In one case, the USAF wanted the F-16 to be compatible with another contractor's new missile, which was delayed by nine months. Meanwhile, the F-16 upgrade process was expected to proceed anyway and finish by the original deadline, despite the additional risk created by the delayed input. These types of situations motivated Lockheed Martin to seek a way to

Figure 7.11.1
F-16 cockpit and avionics system controls (courtesy of Lockheed Martin).

pinpoint the implications of a holdup of an external input and justify any appropriate changes in schedules or other expectations.

Data Collection

Tyson Browning, then an employee at Lockheed Martin, advised a small team on representing the existing process in a DSM. Because of limited resources for building the model, the team used existing process documentation as the basis for an initial DSM, which was then revised slightly through some brief discussions and a meeting with the project manager. Rather than seeking to capture the process completely, model building ended once the model had achieved sufficient richness and accuracy to serve its immediate purpose.

Model

The DSM shown in figure 7.11.2 contains a set of high-level activities constituting the avionics upgrade development process. The activities were originally grouped by organization or specialization, and these designations are indicated in the letters preceding each activity name (such as S for software development, F for flight test, and L for logistics) and by the color coding of the row labels. Note that the DSM actually shows a subset of the development activities because not all activities were relevant in this particular instance (which is why the activity numbers go up to 42 although the DSM does not contain 42 activities). The initial DSM (not shown) was resequenced to arrive at the upper triangular DSM shown here. This reordering of the rows and columns of the matrix inter-mixed activities from the various organizations and moved some activities (such as 29) far upstream in the process. The region above the main DSM shows where some of the external inputs enter the process. We used the IC/FBD convention so that these external inputs would line up with the columns of the activities receiving them.

Results

The input of particular interest, CFE, is used by activity 16, Detailed Design. Activity 16 should occur between activities 15 and 17 (even in the resequenced DSM), but if all of its inputs are not available, then it is faced with two options: (1) wait for the inputs, or (2) proceed based on assumptions about the inputs.

Option one is demonstrated in the DSM by moving activity 16 downstream in the process to the actual point in time where the CFE input shows up. This delay has several consequences. First, the output from activity 16 (that becomes an input to 17) will in turn be delayed, so it now appears below the diagonal in the DSM. Hence, activity 17 must confront the same dilemma faced by activity 16: wait or guess. Activities 25 and 37, in turn,

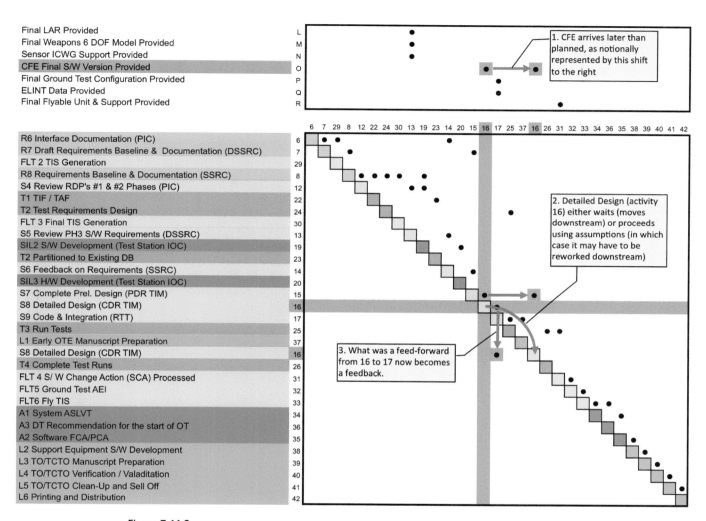

Figure 7.11.2
DSM representation (IC/FBD) of the F-16 software upgrade development process, with external inputs shown above.

must then also face the same dilemma. Hence, any holdup of an activity may cause a cascade of delays through the project, and the DSM allows one to determine the specific activities affected.

Option two, starting the activity at its originally scheduled time on the basis of assumptions about its missing input, may be possible but usually incurs risk; if the assumptions turn out wrong, then the activity must be reworked once the real input arrives. Furthermore, any downstream activities that proceeded based on what may turn out to be a faulty output from the activity may also have to be reworked, and so on. The scope of this cascade of changes can be visualized in the DSM as similar to the effects of delaying the activity (option one). By moving activity 16 downstream in the DSM, until the point where the actual inputs arrive (and any assumptions made earlier can be verified), we see how a mark moves below the diagonal of the matrix. As the distance of this mark from the diagonal grows, so does the scope of the potential cascade of rework.

Lockheed Martin used this DSM model of delays in external inputs as a basis for discussion with USAF customer representatives. Lockheed Martin was able to anticipate the impacts on development cost and schedule, communicate the implications of these delays, and renegotiate specific commitments with other activities in the process in order to adapt in dynamic situations. In particular, it became clearer how external events caused specific internal rework, and how decisions and situations that caused additional marks to move below the diagonal in the DSM increased a project's risks.

Reference

Browning, Tyson R. 2000, September 18. *Notional, Project Risk Management Using the DSM.* Proceedings of the 2nd MIT Design Structure Matrix Workshop, Cambridge, MA.

Example 7.12   Ford Motor Company Hood Development Process

Contributors

Tony P. Zambito
Ford Motor Company

Daniel E. Whitney
Massachusetts Institute of Technology

Ali A. Yassine
American University of Beirut

Problem Statement

Ford Motor Company is one of the world's largest automobile manufacturers. By the 1990s, Ford and other automotive OEMs were faced with unprecedented global competition where product refresh rate and body styling were widely recognized as sources of competitive advantage (see figure 7.12.1). This required faster and more reliable product development processes. The goal of this DSM application was to understand the feasibility and effectiveness of using the DSM to improve a real-life, highly evolved, and iterative process. Improvement was defined in terms of:

Figure 7.12.1
Modern automobile designs include highly styled sheet metal body panels (2010 Taurus, courtesy of Ford Motor Co.).

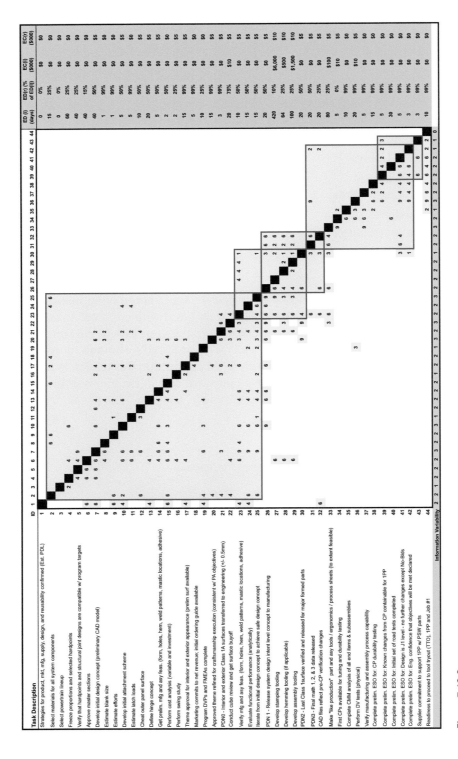

Figure 7.12.2
Baseline DSM, representing the existing hood development process. Annotations indicate how duration, cost, and dependency data were captured in this format.

- Reduction of product development lead time
- Reduction of product development lead-time variation

Data Collection

Tony Zambito, an experienced Ford engineer and master's student in MIT's System Design and Management program, studied the development process for hood design and created a process DSM over a period of approximately nine months in 1999 while executing the existing hood development process. Data for the hood DSM came from three primary sources:

- Interviews with approximately 15 experienced engineers, technical specialists, and managers from styling, engineering, manufacturing, and assembly functions
- Existing development data for task durations, process lead times, nominal resource levels, and typical areas of rework
- Real-time data gathered during the execution of current hood development projects

Model

The DSM model shown in figure 7.12.2 represents Ford's baseline (as-is) hood development process in 1999, as confirmed by interviewees.

Interviewee input and data were used to estimate the cost and duration of initially executing and reworking each activity. These data are shown in the columns at the right of the DSM, where ED(i) and ED(r) represent the initial and rework durations, respectively. Similarly, EC(i) and EC(r) represent initial and rework costs, respectively.

The blocks in the DSM highlight iterative groups of activities in the as-is execution sequence. The blocks were initially identified by inspection using the above-diagonal marks as a guide and then refined through further interviews.

Dependencies between tasks are marked with a numerical index termed "task volatility" (TV), which represents the probability of rework. Specifically, TV is the product of the variability of the input information termed "information variability" (IV) and the receiving task's sensitivity to change in that information, termed "task sensitivity" (TS). That is, TV = IV x TS. These metrics are similar to those used by Krishnan et al. (1997).

Each task was assigned one of three IV levels (shown along the bottom of the DSM in figure 7.12.2) and each dependent task was assigned one of three TS levels (captured in a separate matrix). Figure 7.12.3 describes these levels and the range of possible TV values.

Lead time of this baseline process was simulated using a Monte Carlo simulation (Browning and Eppinger, 2002; example 7.6). This simulation model was calibrated against actual lead times by scaling the task volatility values (Yassine et al. 2001).

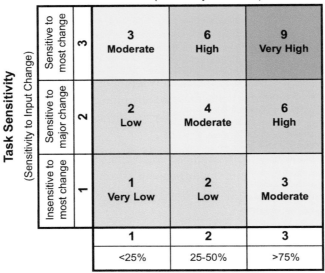

Task Volatility Values
(Probability of Rework)

Task Sensitivity
(Sensitivity to Input Change)

Information Variability
(Likelihood of Input Changing)

Figure 7.12.3
Nine possible task volatility values result from information variability (IV) and task sensitivity (TS) combinations.

Results

The simulation analysis indicated that the mean lead time of the baseline process was 929 days, with a standard deviation of 149 days. This suggests that rework accounts for a significant portion of the lead time and creates substantial variance. Resequencing the matrix using a standard DSM sequencing algorithm only moved one task and consequently showed no lead-time improvement in the simulation.

Reducing lead time therefore required restructuring the dependencies to reduce iteration, which was accomplished by redefining some tasks and methods. For example, there is a dense set of dependencies between task 7, a CAD designer developing the initial CAD model, and task 24, an engineer doing an analysis to evaluate performance of the design. Failure to meet requirements in task 24 results in reworking the design in task 7. The process from task 7 to 24 takes 55 days initially and 28 days for each iteration. This is illustrated in figure 7.12.4.

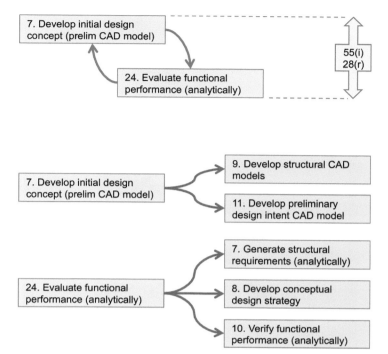

Figure 7.12.4
An example where rework for two coupled tasks required 28 days per iteration. These two tasks were decomposed into five tasks that changed the dependency structure and reduced the iteration time.

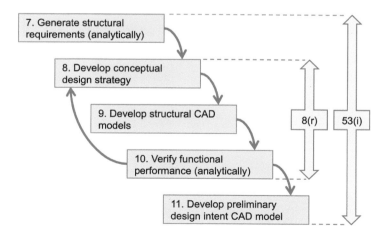

Figure 7.12.5
The restructured tasks removed some of the work from the iterative loop. The entire set of tasks required about the same duration as the original two tasks, but the iteration time was reduced from 28 days to 8 days per iteration.

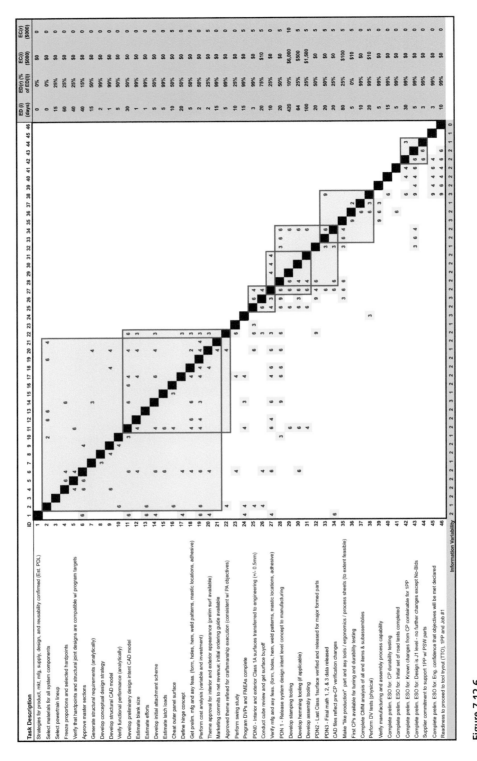

Figure 7.12.6
DSM for the improved hood development process in which several iterative loops were restructured to reduce overall iteration time.

Historical data indicated that a designer typically spends four to five months in design (task 7). This suggests roughly four iterations. DSM simulation estimated that task 7 would be iterated 4.1 times.

Improving this critical iteration loop involved redefining the development process to accelerate the necessary iterations and decouple as many of the other tasks as possible. To do so, we decomposed the two tasks into five, as shown in figure 7.12.4. As illustrated in figure 7.12.5, this change reduced each iteration by 20 days. Figure 7.12.6 shows the resulting DSM and the effect on the design iteration loop.

The result of this DSM application was a dramatic reduction in hood development lead time (and variance), from 929 (149) days to 772 (43) days, according to the simulation. The first hood developed using this process took 790 days and also technically outperformed the last hood developed under the baseline process.

The benefit of reducing timing variance is often overlooked. However, variance reduction gives the program manager confidence in the schedule and thus confidence to perform extra iterations where they can be helpful for performance. This is somewhat counterintuitive, as the common understanding is that iterations are bad. Even experienced project analysts generally compute only the mean duration and do not expect that estimating the variance will add any special insight. Nevertheless, uncertainty and risk reduction are primary considerations in complex product development projects, so analysis of variance should receive more attention.

References

Browning, Tyson R., and Steven D. Eppinger. 2002, November. Modeling Impacts of Process Architecture on Cost and Schedule Risk in Product Development. *IEEE Transactions on Engineering Management* 49 (4):428–442.

Krishnan, Viswanathan, Steven D. Eppinger, and Daniel E. Whitney. 1997, April. A Model-Based Framework to Overlap Product Development Activities. *Management Science* 43 (4):437–451.

Yassine, Ali A., Daniel E. Whitney, and Tony P. Zambito. 2001. *Assessment of Rework Probabilities for Simulating Product Development Processes Using the Design Structure Matrix (DSM)*. ASME Design Engineering Technical Conferences, DTM-21693.

Example 7.13   Alfa Laval AB Heat Exchanger Design

Contributors

Ingvar Rask
SSPA

Staffan Sunnersjö
School of Engineering, Jönköping University

Both authors were previously at the Swedish Institute of Production Research, IVF, at the time of this example.

Problem Statement

Alfa Laval AB is a manufacturing company active in about 100 countries and supplying systems for liquid separation, heat treatment, and fluid handling. One of its product lines is a range of plate heat exchangers for use in process industries such as food and energy production. One example is shown in figure 7.13.1. The heat exchanger consists of a stack of plates each with pressed channels that contain process fluids. The patterns and dimensions of the channels are of critical importance for the system's performance. Because new variants of the basic concepts are frequently required, the company decided to develop a computerized system for automated channel design. For this purpose, the network of tasks and their dependencies in the design process were analyzed using a process architecture DSM.

Data Collection

Clarifying the steps of the design process was a project lasting about three years and was done in parallel with development of the design automation system. The work was done by Ingvar Rask in cooperation with company experts. The design process includes engineering design, stress analysis, fluid flow analysis, and manufacturing processes. Thus, company expertise from several disciplines was captured in the design process tasks and their interdependencies. During this project, the need for a structured approach gradually emerged, and it was decided to employ a DSM.

When the prototype design automation system was tested and the development of the final system was approved, the structure of the design process was clearly represented by the DSM models.

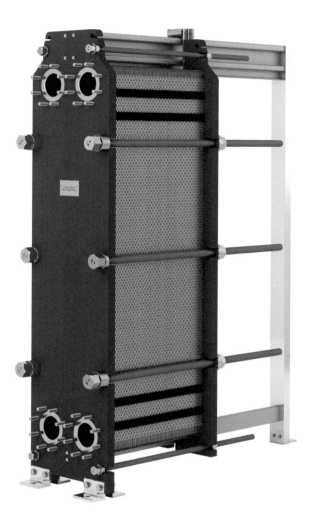

Figure 7.13.1
Plate heat exchanger (courtesy of Alfa Laval Lund AB).

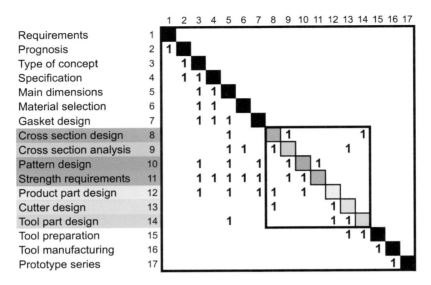

Figure 7.13.2
Sequenced DSM identifying a block of seven coupled tasks in the heat plate exchanger design process.

Model

The main steps of the design process are represented in the DSM shown in figure 7.13.2. The tasks are represented in a sequence that has evolved over many years of practical experience.

The DSM identifies seven coupled tasks forming a circuit involving design and analysis of flow channel cross section (8, 9), channel pattern design (10), strength analysis (11), plate design (12), design of the cutter for machining the press tool (13), and press tool design (14). These tasks require involvement from three different departments: design office, strength analysis, and production preparation.

To analyze the circuit in enough detail to devise a solution algorithm, tasks 8–14 were expanded to the parameter level. The resulting matrix, before analysis, is given in figure 7.13.3.

Using a standard algorithm, we resequenced the DSM to arrive at the DSM in figure 7.13.4, where it is apparent that the dependencies form two clusters of parameters that relate to two separable design issues—design of flow channel cross section and flow channel pattern (layout).

Reaching a solution requires an iterative approach with assumed starting values for parameters with feedback dependencies (the seven superdiagonal marks remaining in the DSM). This is in fact an optimization problem where the coupled parameters are best solved using a suitable standard optimization algorithm.

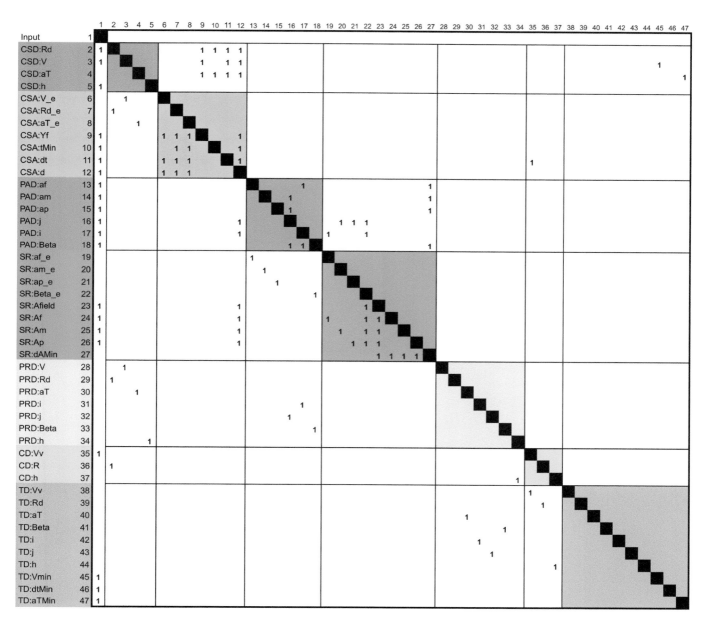

Figure 7.13.3
Expansion of the coupled block of seven design tasks to the parameter level.

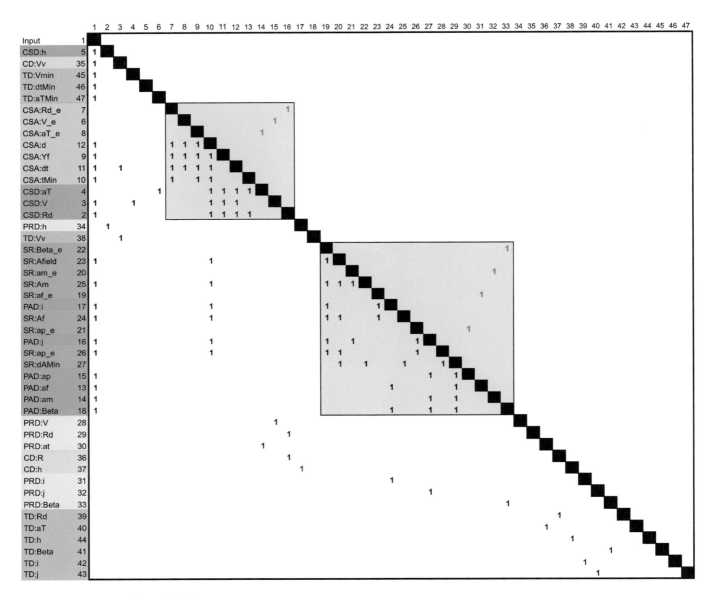

Figure 7.13.4
Resequenced parameter-level DSM intermingling the original seven design tasks.

Results

It is a common experience that when a work task is to be computerized, the manual process that it supports or replaces needs to be described in a much more exhaustive and stringent way than was previously necessary. This might be an obstacle, but it also provides a great opportunity to review and streamline the process.

This application demonstrated how the DSM model and analysis can reveal critical characteristics of the design problem. This showed how to implement the design automation system in an optimal way. The DSM identified the sequence of process steps, including where parallel work can be executed and how the coupled tasks could best be solved through iterations. This is a natural extension of the use of DSMs for process mapping and planning and satisfies the increased stringency required when developing design automation systems.

References

Amen, Rafael, Ingvar Rask, and Staffan Sunnersjö. 1999, September 12–15. *Matching Design Tasks to Knowledge-Based Software Tools—When Intuition Does Not Suffice*. Proceedings of the ASME International Design Engineering Technical Conferences (Design Theory & Methodology Conference), Las Vegas, NV.

Rask, Ingvar, and Staffan Sunnersjö. 1998, October 6–8. *Design Structure Matrices for the Planning of Rule-Based Engineering Systems*. Proceedings of the European Conference on Integration in Manufacturing, Göteborg, Sweden.

Example 7.14   Elevator Design Process

Contributor

Sule Tasli Pektas
Bilkent University

Problem Statement

The Industry Foundation Classes (IFC) project is the world's largest effort to date aimed at standardizing the representation of building product and process knowledge. IFCs are developed by an international nonprofit organization named BuildingSMART and have become widely accepted as the international standard. The process modeling methods used in the IFC development are IDEF0 and Business Process Modeling Notation (BPMN). This study observed two important limitations of the IFC process modeling:

- IDEF0 and BPMN are only able to create well-structured models when the activities include a sufficient level of detail. They represent the dependencies in the process in a limited way, so it is difficult to see the true architecture of the process.

- The tools were employed in merely a top-down fashion, where the modeling begins at a high level and is decomposed as needed. However, it is also useful to go backward (i.e., to use the deliverables as building blocks and integrate the model from the bottom-up). This also helps to verify the accuracy of the interactions in the model.

Thus, this study demonstrated the complementary use of parameter-based DSM models with conventional higher level process models in the construction industry.

Data Collection

We applied the parameter-based DSM modeling in a case study of elevator design. An architectural office, its engineering collaborators, and an elevator provider participated in the study. Sule Tasli Pektas collected the data through inspection of design documents and interviews with designers over five months.

First, higher level IDEF0 process models of the elevator design process were produced in compliance with the IFC process modeling notation. Then a parameter-based DSM model of the process was developed to provide better insights into the processes. The data collection process was highly iterative; the draft models were often revised according to the comments received from the participants.

Model

In the DSM model shown in figure 7.14.1, marks in a row represent inputs to a parameter decision while marks in a column represent the output results of the parameter decision (IR/FAD convention). Colors associate the parameter decisions with higher level activities. Parameters highlighted in red on the diagonal are critical ones that would appear to cause large iteration cycles in the process.

Results

This bottom-up, parameter-based approach provided new insights into the higher level tasks and allowed the improved process to be based on the rational and natural information flows rather than superficial assumptions. To illustrate how the parameter-based DSM helped to improve the higher level models in the case study, a simple example was extracted from the large models.

Figure 7.14.2 shows two coupled activities in the elevator design process. However, the detailed structure within this cycle (i.e., which parameter decisions within the activities depend on each other) is not clear from either the IDEF0 view (a) or from the high-level activity-based DSM (b).

However, the parameter-based DSM decomposes the two-coupled activities to the parameter level. This shows the parameter decisions in a more detailed process map (figure 7.14.3a). When this DSM is resequenced, the appropriate ordering of decisions is obtained, and, in this case, the iteration is removed from the process (figure 7.14.3b). As a result, the parameters in the process can be regrouped into three activities instead of the original two. In this way, the integrated process model can be based on the more detailed information flows rather than just the overview activities.

Of course, this example is simple, and in many cases iteration cannot be totally removed from engineering design processes. However, we applied this approach in two case studies in building design, and we believe that the findings of these studies supported the complementary uses of the parameter-based DSM with the conventional IFC process models.

One challenge of the parameter-based DSM observed in this study is the large number of parameters to be determined by the design processes. However, capturing and managing all parameter decisions in a process may not be necessary. In order to increase the efficiency of the models, the parameter-based DSM decomposition can be used only for the problematic activities such as highly coupled activities, activities that involve many actors, or critical activities that tend to cause delays in the process. Thus, this study demonstrated the functionality of the parameter-based DSM. We believe that this procedure can be further explored and exploited in many ways.

	1	2	3	4	5	6	7	9	8	10	11	12	13	14	15	16	17	18	19	20	21	22	23	24	25	26	27	28	29	31	32	33	34	35	36	37	38
Building Type — 1	**1**																																				
Building Style — 2		**2**																																			
Tenancy Type — 3	X		**3**																																		
Floor Area — 4				**4**																																X	X
Building Structure Layout — 5					**5**																											X				X	X
Number of Floors Served above Main Terminal — 6						**6**																															
Average Interfloor Distance — 7							**7**																														
Passenger Arrival Rate — 9	X							**9**																													
Building Population — 8	X	X	X	X					**8**																												
Uppeak Interval — 10	X									**10**																X		X									
Average Number of Passengers per Trip — 11										X	**11**	X																									
Contract Capacity — 12	X										X	**12**																									
Average Highest Call Reversal Floor — 13							X				X	X	**13**																								
Average Number of Stops — 14												X		**14**																							
Contract Speed — 15	X						X					X			**15**							X															
Single Floor Transit Time — 16	X					X						X			X	**16**																					
Car Door Opening Configuration — 17												X					**17**																				
Door Opening/Closing Time — 18																	X	**18**																			
Time Consumed when Stopping — 19																X			**19**		X																
With Gearing — 20																X				**20**																	
Floor Cycle Time — 21																	X		X		**21**																
Average Passenger Transfer Time — 22	X																X					**22**						X			X	X					
Round Trip Time — 23											X		X	X			X				X		**23**														
Uppeak Handling Capacity — 24										X	X												X	**24**	X												
Number of Elevators — 25											X		X										X	X	**25**			X	X							X	X
Car Door Width — 26						X						X						X								**26**											
Elevator Type — 27					X													X									**27**										
Car Width — 28													X													X	X	**28**							X		
Car Depth — 29													X													X			**29**								X
Car Grouping — 31			X	X																							X	X		**31**							
Structural Frame LH Side Clearance — 32																															**32**						

Figure 7.14.1
A partial view of the partitioned DSM model of the elevator design process.

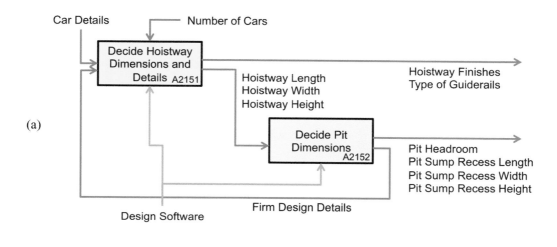

Figure 7.14.2
Two coupled activities in the elevator design process: IDEF0 view (a) and activity-based DSM view (b).

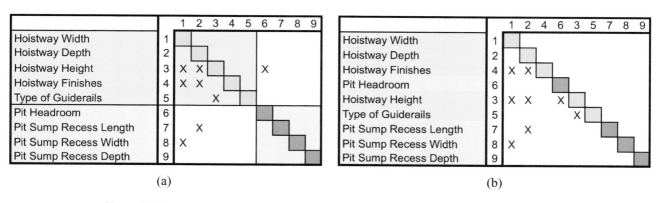

Figure 7.14.3
Two coupled activities of the elevator design process: the initial parameter-based DSM (a) and the resequenced DSM (b).

References

Pektas, Sule T. 2010, July. *The Complementary Use of the Parameter-Based Design Structure Matrix and the IFC Process Models for Integration in the Construction Industry*. Proceedings of the 12th International Dependency Structure Modelling (DSM) Conference: Managing Complexity by Modelling Dependencies, Cambridge, UK, pp. 389–402.

Pektas, Sule T., and Mustafa Pultar. 2006. Modelling Detailed Information Flows in Building Design with the Parameter-based Design Structure Matrix. *Design Studies* 27 (1):99–122.

Example 7.15   L.L.Bean Software Code Base

Contributors

Carl Hinsman
L.L.Bean, Inc.

Neeraj Sangal
Lattix, Inc.

Judith Stafford
Tufts University

Problem Statement

L.L.Bean is a large retail business whose development processes must be agile in order to allow rapid enhancement and maintenance of its technology infrastructure. Over the past decade, L.L.Bean's software code base had become brittle and difficult to maintain. An effort was launched to identify and develop new approaches to software development that would enable ongoing agility to support the ever-increasing demands of a successful business. This example summarizes L.L.Bean's effort to restructure its code base and adopt process improvements that support an agile, architecture-based approach to software development, governance, and maintenance.

Data Collection

Over a period of six months in 2006–2007, a small team of software engineers at L.L.Bean undertook two key tasks. First, we researched the abstract nature of software architecture primarily through articles and academic papers. Managing dependencies is not a new problem, and considerable research and analysis on a wide range of approaches was available. Second, we created a detailed model of the existing static dependencies in L.L.Bean's Java code base and identified patterns in those dependencies.

Initially, a combination of open-source tools was used for graphing and dependency analysis. The processes were computationally intensive, and there was a limit to the amount of code that could be analyzed collectively. Furthermore, these views were often incomprehensible and of little practical value in either communicating or managing the architecture. L.L.Bean's research identified the Lattix DSM-based dependency analysis tool as promising. The Lattix tool offered a comprehensive and easy-to-understand user interface, a mechanism for prototyping and applying architecture rules, and support for "what if" analysis without code modification.

Model

The DSM model was created from L.L.Bean's Java-based application software. A Java compiler generates Java byte code by compiling Java source code into Java classes. These Java classes are in turn aggregated into Jar files. The Jar files were loaded into Lattix, which extracted the interdependencies and created the DSM model. For purposes of this model, we loaded Jar files for the infrastructure and the various application domains into Lattix. The model was constructed by loading in more than 100 Jar files for a system of more than one million lines of code. The code is organized into more than 100 packages composed of 3,000 classes, thus yielding a DSM (figure 7.15.1) clearly illustrating all dependencies within the modeled code base. We accounted for dependencies such as invocations, inheritances, data member references, and constructs.

Lattix generates a hierarchical DSM that is initially organized by Jar files and then by the package and class structure within each Jar file. At each level of the hierarchy, it is easy to see the coupling by applying a partitioning (sequencing) algorithm. The initial model gave us a big-picture view of how various parts of the software were coupled together. The numbers in the off-diagonal cells of the DSM indicate the strength of dependency from one component to another.

Results

The initial model was a hierarchical DSM that reflects the Jar files and the package/class hierarchy. This model was then transformed to the desired architecture. This meant creating abstractions for different layers, domains, and applications. The Java class components were grouped into three categories: *domain-independent*, *domain-specific*, and *application-specific*. These classes were organized into common layers according to their specificity, with the most generalized layers at the bottom and the most specific layers at the top. In this approach, each layer is governed by the principle that members of a given layer can only depend on other members in the same layer or in layers below it. Each layer, or smaller subset within a layer, is assembled in a cohesive unit, often referred to as a program library or subsystem. These cohesive units were Java Jar files. This approach produced a set of independently developable components that are not coupled by cyclical interdependencies.

Modeling the desired architecture also gave us visibility into undesirable dependencies. Three key errors were identified that were at the core of the interdependency entanglement. We classified these as *Misplaced Common Types*, *Misplaced Inheritable Concrete Classes*, and *Catch-all Subsystems*. Within Lattix, we moved Java classes to their appropriate package according to both their generality/specificity and their behavior/responsibility. At the end of this process, we were surprised to observe that nearly all undesirable dependencies at the top level had been eliminated. (Note the absence of superdiagonal

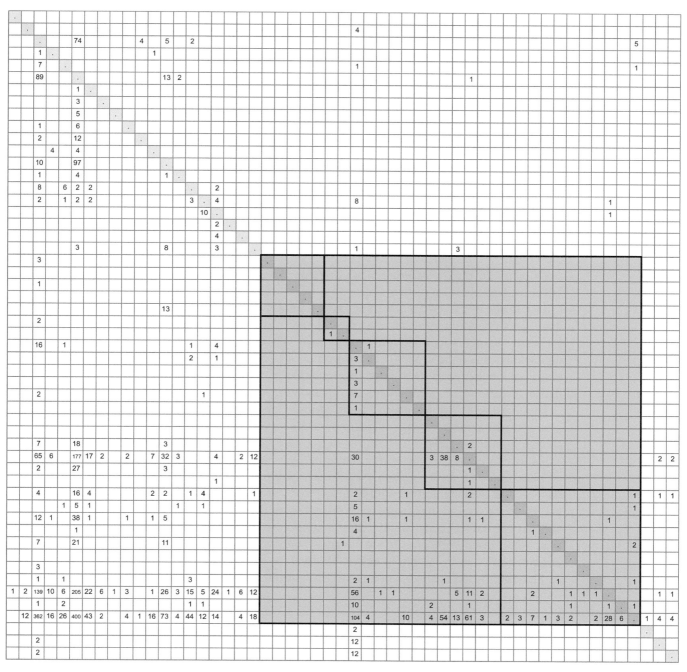

Figure 7.15.1
An excerpt from the full DSM, showing the partitioning result for a selected module.

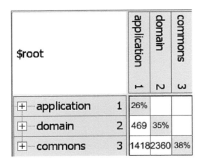

$root		application 1	domain 2	commons 3
⊞ application	1	26%		
⊞ domain	2	469	35%	
⊞ commons	3	1418	2360	38%

Figure 7.15.2
Top-level DSM showing the total number of dependencies among the three main modules.

entries in figure 7.15.2.) Furthermore, modeling the desired architecture also gave us visibility into the causes for the cyclical dependencies between Jar files that had caused difficulties in our build system.

To maintain the architecture, a set of rules was created that could be applied to the DSM models, thereby further improving the visibility of maintenance processes. These rules enforce a layered architecture and essentially state that members of a given layer may only depend on other members in the same layer or in lower layers. Rules also help software engineers identify reuse candidates. When violations occur, the nature of the dependencies and the specific behavior of the Java code are closely analyzed. If there are multiple dependencies on a single resource that break an allowed dependency rule, then the target resource is a candidate for repackaging to a lower level.

The DSM model provided consistent visibility and supported ongoing communication among development teams, configuration engineers, and project leaders. It also facilitated change impact analysis.

L.L.Bean found that increasing the visibility of software architecture greatly reduced architectural drift as the system evolved and at the same time reduced ongoing maintenance costs. Architectural visibility also provided guidance for large-scale refactoring. L.L.Bean discovered that changing the structure of the system can sometimes be achieved without substantial code modification. Large-scale reorganization is a complex process that, when done with proper tool support and in a disciplined software development environment, can be highly effective. The results of this experience demonstrated that architecture-based analysis can improve the productivity of software development.

References

Clements, P., F. Bachmann, L. Bass, D. Garlan, J. Ivers, R. Little, R. Nord, and J. Stafford. 2003. *Documenting Software Architectures: Views and Beyond*. New York: Addison Wesley.

Although complex engineering projects may be considered to be individual complex systems, a number of different systems have been identified, modeled, and studied within projects. So far this book has focused on three such systems: the desired result (product), the work done to get to that result (process), and the people who do the work (organization). Maintaining the distinctions between these systems has enabled focused modeling and the generation of insights that might not have been as apparent otherwise. For example, Browning et al. (2006) distinguished five critical domains in a project (figure 8.2): the product system (desired result), the process system (activities done to get the product), the organization system (organizational units that perform activities), the tool system (tools, technologies, facilities, and resources used by people to do activities), and the goal system (requirements, targets, objectives, and constraints for and on the other four domains). In complex projects, each of these domains is a complex system, each has an architecture, and each affects the others. (Moreover, as figure 8.2 indicates, these domains may also interact across projects, such as when an enterprise endeavors to use a common process, tool set, or organizational resources across projects.) To explore such cross-domain effects, modelers need multidomain methods. Because the DSM has shown great benefits for modeling and gaining insight into complex systems, it is not surprising that extensions to the basic DSM have developed to enable such efforts.

In this chapter, we discuss three types of cross-domain modeling constructs related to the DSM:

1. The *1.5-domain DSM* (1.5d DSM) extends the basic DSM by adding an enhanced representation scheme (such as color coding) to project the shadow of one domain

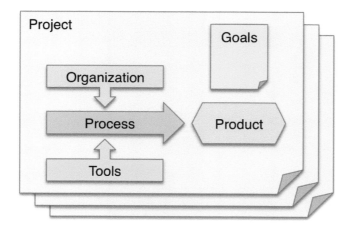

Figure 8.2
Five domains or systems in a project (adapted from Browning et al. 2006).

(such as the organizational unit responsible for an activity) onto the DSM of a focal domain (such as the activities in a process).

2. The *domain mapping matrix* (DMM) is a rectangular matrix that shows the relationships between two domains (such as people assigned to activities). A DMM does not show the relationships within either of the domains; it only shows the mapping between them.

3. The *multidomain matrix* (MDM) combines two or more DSMs and DMMs into a larger, multisystem (or "system of systems") model.

Extending DSM to More Than One Domain: 1.5d DSMs

A simple extension of the common single-domain DSM model is achieved by labeling the DSM elements according to their relationship with elements in a secondary domain. This can be implemented by adding one (or more) columns next to the element labels, indicating their situation in the second domain. Figure 8.3 shows a simple example of the 1.5d DSM approach. In this illustration, there are nine DSM elements (labeled 1–9) in the primary domain. They are mapped to three elements (labeled A, B, and C) in the secondary domain. The DSM is partitioned based on the structure of the primary domain. Colored shading of the names according to the secondary domain assists in understanding the cross-domain mapping.

Typical uses of a 1.5d DSM would be to tag the tasks in a process DSM with the organizational responsibility of each task (see example 7.4) or to identify the suppliers of each component represented in a product DSM (see example 9.3).

Secondary	Primary	1	2	3	4	5	6	7	8	9
A	1	1								
B	2	X	2							
A	3		X	3						
C	4	X	X		4					
A	5			X	X	5	X		X	
B	6	X	X		X	X	6			
B	7	X			X			7	X	
C	8		X				X	X	8	
A	9				X			X		9

Figure 8.3
The 1.5d DSM represents a primary DSM domain and labels the elements with a second domain.

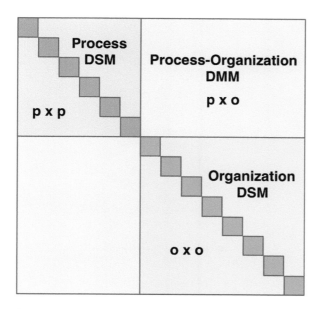

Figure 8.4
The DMM relates the elements of one DSM domain to elements of another DSM domain—in this case, process activities to organizational units.

Mapping Between Two Domains Using the DMM

Rectangular matrices are commonly used to map the relationships between two sets of items. Danilovic and Browning (2007) reviewed several examples of such uses and codified the term DMM as an inclusive term, complementary to DSM. The DMM is a rectangular ($n \times m$) matrix mapping between two domains, such as the process and organization domains shown in figure 8.4. Each individual domain may be modeled with a DSM, which captures the internal relationships between its elements (and sometimes also external relationships with elements of the same type) but not relationships to the elements in other domains. Like a DSM, a DMM may be binary, merely indicating the presence or absence of a direct relationship, or it may contain numbers or other symbols indicating the strength, degree, or type of relationship across domains. For example, the DMM in figure 8.4 could correspond to the responsibility allocation matrix (RAM) used by project managers, which is also called a RACI chart because it can be used to indicate four types of person-to-activity relationships: responsible, accountable, consult, and inform (PMI 2008).

Figure 8.5 shows a binary DMM mapping customer requirements to product specifications (i.e., essentially a top-level QFD matrix). Figure 8.6 shows the same data after a clustering analysis, identifying four major and one minor cluster with the rectangular

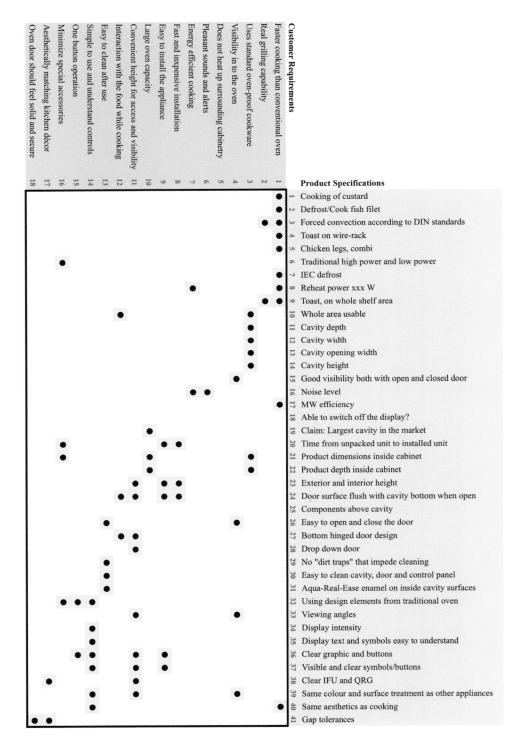

Figure 8.5
Example of a DMM before clustering analysis (adapted from Danilovic and Browning 2007).

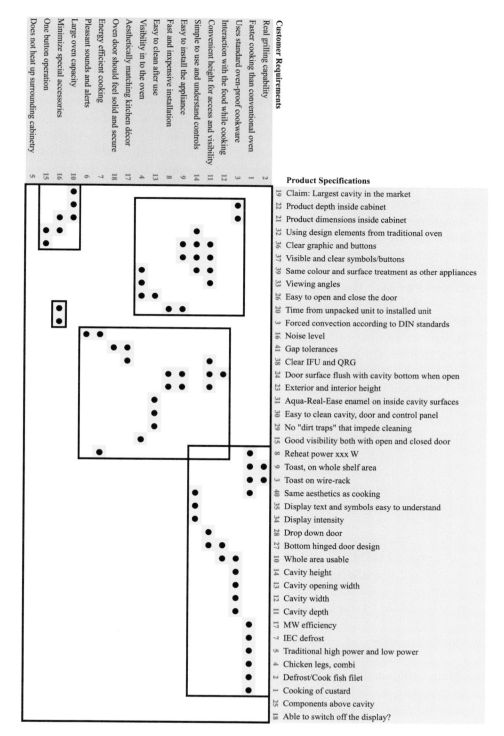

Figure 8.6
Example of a DMM after clustering analysis (adapted from Danilovic and Browning 2007).

outlines. (Note that the DMM does not necessarily contain marks along a diagonal, so the clusters may appear anywhere in the matrix.) This analysis shows how a particular group of customer requirements is addressed by a set of product specifications. (In traditional QFD analysis, this tends to happen only in terms of individual elements, not sets thereof.) The clusters indicate where a high level of relationships requires intense coordination across domains or within each domain due to mutual relationships with the other. Figure 8.6 also identifies an area without relationships between the domains, the last two columns, where two product specifications do not correspond to any customer requirements. This implies that we might be missing some important information from the customer requirements, or that we might have introduced some superfluous product specifications. Thus, DMMs can help clarify the relationships between domains, and they may furthermore help verify the elements comprising each domain.

Modeling Within and Between Two or More Domains Using the MDM

DSMs and DMMs may be used in conjunction to analyze the influences of one system domain on another or to infer the presence of elements and relationships in one system from another. This possibility led to the proposition of a matrix of matrices, as illustrated in figure 8.7. When Mendeleev proposed the periodic table of elements, not all of the elements had been discovered, but their likelihood of existing could be inferred from the open slots in the table. Similarly, at the time of figure 8.7 (circa 2004), not all of the DSMs and DMMs existed in actual applications. Since then, various combinations of DSMs and DMMs have been proposed and explored, including subdomains of the product system (Danilovic and Browning 2007), the Engineering Systems Matrix (Bartolomei 2007), and several other combinations (e.g., Lindemann et al. 2009; Maurer 2007). By 2007, Maurer had codified the term multidomain matrix to refer to such applications, and this term gained popularity in the DSM community.

Analysis techniques for the MDM are still being contemplated and developed. It is not yet clear how best to analyze an MDM holistically because it contains a mixture of static and temporal DSMs. Should an MDM be clustered, sequenced, or both? Or might some other kind of analysis provide still further insights? Several of the presentations at the recent DSM conferences provide further explorations of these possibilities and applications, as do the examples in chapter 9.

Special Case for Two Domains

When two domains are decomposed such that there is a one-to-one mapping from one to the other, we have a special case that can be considered without using a DMM. Several possible examples may be: One person is assigned to each process activity, one team is designated for each product component, or one product specification corresponds to each

Goals DSM $g \times g$	Goals-Product DMM $g \times d$	Goals-Process DMM $g \times p$	Goals-Organization DMM $g \times o$	Goals-Tools DMM $g \times t$
	Product DSM $d \times d$	Product-Process DMM $d \times p$	Product-Organization DMM $d \times o$	Product-Tools DMM $d \times t$
		Process DSM $p \times p$	Process-Organization DMM $p \times o$	Process-Tools DMM $p \times t$
			Organization DSM $o \times o$	Org.-Tools DMM $o \times t$
				Tools DSM $t \times t$

Figure 8.7
"Periodic table" of DSMs and DMMs, forming an MDM (adapted from Danilovic and Browning 2007).

customer need. The one-to-one mapping means that both DSM domains have the same number of elements and that the DMM relating them would be trivial (an identity matrix). Moreover, we can directly compare the two DSMs by ordering the elements the same way in both matrices and examining the set of off-diagonal interactions in both matrices. Analysis of this sort has compared product architecture to organizational architecture, with fascinating insights regarding organizational effectiveness (see example 9.2).

Applying DMMs and MDMs

DMMs and MDMs have been applied to a range of industrial problems and have begun to produce many useful insights. Many examples are given in the next chapter. Typical and potential applications include:

- Identifying needs for cross-functional, cross-team interactions in an organization based on interactions among product components or process activities (see examples 9.2, 9.4, 9.6, 9.12).

- Inferring elements and/or interactions in other domains. A DSM and a DMM can be used to infer the intra-domain relationships in another DSM. This can be used both as a starting point for building another DSM and as a means of verifying the information therein. Furthermore, in a dynamic sense, changes in one domain may signal or trigger particular changes in the other, making an MDM a potential source of leading indicators of product, process, organizational, or other changes. Such inferences can also be used to focus the attention of those modeling or monitoring the system on particular areas of expected interest (see examples 9.5, 9.6, 9.11, 9.12).

- Project architecting and evolution. How do the domains comprising a project affect each other? How do similar domains relate across projects? Is there a preferable order to the design of the systems comprising a project? That is, perhaps the desired result (product architecture) should determine the appropriate activities to be done (process architecture), which should then determine the appropriate organization architecture. But then what process and organization architectures should be used to determine the product architecture? Clearly, each of these systems must co-evolve over the course of a project. An MDM can help model and analyze these dynamics and the emergent behaviors of the systems (see examples 9.6, 9.7, 9.11, 9.12).

Conclusion

MDMs provide a promising avenue for modeling complex, multidomain systems such as projects. MDM models can be rich and complex, encapsulating a lot of information, so they hold great potential for applications that require organizing, managing, and analyzing large amounts of information about product, process, organization, and other elements and their intra- and interdomain relationships.

References

Morelli's master's thesis at MIT, summarized in this *IEEE Transactions* article, developed the first multidomain DSM model, with a mapping between and comparison of process to organization domains.

Morelli, Mark D., Steven D. Eppinger, and Rosaline K. Gulati. 1995, August. Predicting Technical Communication in Product Development Organizations. *IEEE Transactions on Engineering Management* 42 (3):215–222.

Eppinger and Salminen presented the concept of using models in the three primary DSM domains and explained how they could be mapped and compared across domains. Browning also discussed several opportunities for applying DSMs across these domains.

Browning, Tyson R. 2001. Applying the Design Structure Matrix to System Decomposition and Integration Problems: A Review and New Directions. *IEEE Transactions on Engineering Management* 48 (3):292–306.

Eppinger, Steven D., and Vesa Salminen. 2001, August. *Patterns of Product Development Interactions*. International Conference on Engineering Design, Glasgow, Scotland.

Gulati and Eppinger explored the co-evolution of the product and organization architectures of complex products, using examples from the automobile industry. Sosa's doctoral thesis at MIT, summarized in a *Management Science* article, utilized DSM models in both domains and compared them to assess the (mis)alignment between architectures across the product and organization domains (see also example 9.2).

Eppinger, Steven D., and Rosaline K. Gulati. 1996, May. *The Coupling of Product Architecture and Organizational Structure Decisions*. MIT Sloan School of Management, Working Paper no. 3906.

Sosa, Manuel E., Steven D. Eppinger, and Craig M. Rowles. 2004, December. The Misalignment of Product Architecture and Organizational Structure in Complex Product Development. *Management Science* 50 (12):1674–1689.

Based on a paper and presentation at the 2004 DSM Conference, Danilovic and Browning formalized the DMM construct and first suggested a "periodic table" array of DSMs and DMMs that would later be called an MDM.

Danilovic, Mike, and Tyson R. Browning. 2007. Managing Complex Product Development Projects with Design Structure Matrices and Domain Mapping Matrices. *International Journal of Project Management* 25 (3):300–314.

Maurer's doctoral dissertation first proposed the term MDM and explored several ways that MDM models can be used.

Maurer, Maik S. *Structural Awareness in Complex Product Design*. 2007. PhD thesis, Technischen Universität München, Munich, Germany.

Lindemann, Udo, Maik Maurer, and Thomas Braun. 2009. *Structural Complexity Management: An Approach for the Field of Product Design*. Berlin, Germany: Springer.

Bartolomei's dissertation investigated the use of MDMs (calling them Engineering Systems Matrices [ESMs]) for modeling sociotechnical systems and the potential use of real options in such systems.

Bartolomei, Jason E. 2007. *Qualitative Knowledge Construction for Engineering Systems: Extending the Design Structure Matrix Methodology in Scope and Procedure*. PhD thesis, Massachusetts Institute of Technology, Cambridge, MA.

Several types of cross-domain mapping matrices have been used in engineering and project management, including Axiomatic Design, Quality Function Deployment, and the Responsibility Allocation Matrix.

Akao, Yoji, ed. 1990. *Quality Function Deployment*. Cambridge, MA: Productivity Press.

Carley, Kathleen M., and David Krackhardt. 1999, June. *A Typology for C2 Measures*. Proceedings of the 1999 International Symposium on Command and Control Research and Technology, Newport, RI.

Krackhardt, David, and Kathleen M. Carley. 1998, June. *A PCANS Model of Structure in Organizations*. Proceedings of the 1998 International Symposium on Command and Control Research and Technology, Monterey, CA.

PMI. 2008. *A Guide to the Project Management Body of Knowledge*. 4th ed. Newtown Square, PA: Project Management Institute.

Suh, Nam P. 2001. *Axiomatic Design*. New York: Oxford University Press.

Browning et al. discussed five interacting domains in complex projects.

Browning, Tyson R., Ernst Fricke, and Herbert Negele. 2006. Key Concepts in Modeling Product Development Processes. *Systems Engineering* 9 (2):104–128.

9 Multidomain Architecture MDM Examples

Overview

This chapter presents 13 example applications of matrix models representing architectures in multiple domains as listed in the table below. Each example describes the purpose of the model (problem to be addressed), how the data were collected, how the model was built, and the results. Where available, references for further information are also provided.

Example	Application	Organization	Purpose
9.1	Hybrid vehicle architecture concepts (MDM)	BMW, Germany	▪ Compare alternative product architectures for hybrid automobiles in terms of structure and functional capabilities
9.2	Jet engine product and organizational structures (two DSMs)	Pratt & Whitney, USA	▪ Explore alignment of architectures in product and organization domains
9.3	Mailing system (1.5d DSM)	Pitney Bowes, USA	▪ Identify opportunities for and impact of component and module design outsourcing
9.4	Team composition for collaboration (DMM)	Audi AG, Germany	▪ Formalize interactions between design and simulation departments
9.5	Political organization (DMM and DSM)	United States Senate, USA	▪ Identify organizational structure of interactions between members, inferred from joint committee assignments
9.6	Multidisciplinary development of electric sunroof (MDM)	BMW, Germany	▪ Provide multidisciplinary system understanding and an effective interlinking of the discipline-specific development processes
9.7	Adhesive anchors and dispensers (MDM)	HILTI, Germany	▪ Visualize developers' interdisciplinary change impact ▪ Identify suitable possibilities for adjusting the system's configuration ▪ Support the design of experiments ▪ Identify measures for better control of the system's complexity

(continued)

Example	Application	Organization	Purpose
9.8	Change packaging in systems design (MDM)	Digital Research Labs, UK	• Assist in identifying the most appropriate change processing approach for a given project
9.9	Airport security (MDM)	Bauhaus Luftfahrt e.V., Germany	• Explore possible future threat scenarios with respect to existing security measures
9.10	Large-scale integrated chip design for a 4G mobile phone (MDM)	Japan Society for the Promotion of Science, Japan	• Find an improved chip design based on better understanding of the initial design processes
9.11	Automobile body-in-white development (MDM)	Audi AG, Germany	• Support a balanced improvement approach, incorporating process, organizational, and information technology aspects
9.12	Miniaturized unmanned air vehicle development (MDM)	Air Force Research Laboratory, USA	• Examine the impact of engineer turnover within the design organization • Examine the effects of changing requirements on the design • Examine design evolution • Identify platform and modularity opportunities • Explore the sources and effects of design changes
9.13	Industrial supply chain network (MDM)	Kalmar Industries, Sweden	• Develop collaboration plan • Design information exchange process

Example 9.1   BMW Hybrid Vehicle Architecture Concepts

Contributor

Carlos Gorbea
BMW and Technische Universität München

Problem Statement

The BMW Group is a leading vehicle manufacturer based in Munich, Germany, known for the BMW and MINI premium performance brands. In 2008, BMW's vehicle architecture division investigated the structural relationships between functions and components of various hybrid vehicle configurations. The study aimed at understanding how hybrid vehicle concepts differ in structure and functional capability based on changes in their basic configuration using MDMs.

Data Collection

The data collection was led by Carlos Gorbea during approximately four months of his doctoral research conducted at BMW's Innovation and Development Center in Munich. The work was performed alongside the BMW Future Hybrids Development Team and Professor Udo Lindemann from the Institute for Product Development at Technische Universität München (TUM). The data to build two-domain MDMs for eight hybrid vehicle powertrain subsystem concepts were collected by means of workshops and meetings with BMW subject matter experts. Subsequent analysis and interpretation of the topic was worked with the help of Dipl. Ing. Tobias Spielmanleitner's thesis work at TUM in 2008.

Model

Each MDM model is composed of a product architecture DSM showing physical component connections (symmetric) and a product function DSM showing input and output energy flows of functional relationships (non-symmetric). A DMM relates these two domains by showing which components provide which functions. (Note that these relationships between functions and components are non-directional, so the placement of the DMM above or below the diagonal of the two DSMs does not matter.)

Each function and component is assigned a unique index number. An example MDM representation for an integrated starter generator (ISG) hybrid powertrain is shown in figure 9.1.1. Each product architecture MDM includes row and column entries for each identified index number—including null components and functions not present within

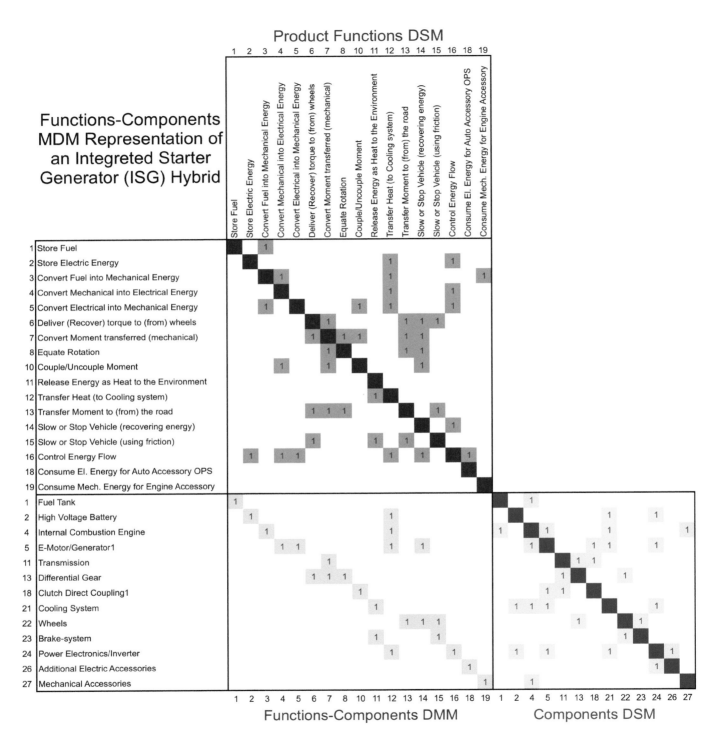

Figure 9.1.1
Function and component MDM for an integrated starter generator (ISG) hybrid powertrain.

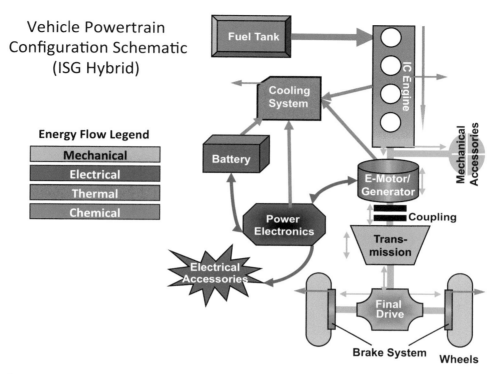

Figure 9.1.2
Schematic depiction used as a guide to the MDM creation.

the architecture—to ensure that matrix size remains the same during matrix manipulations. These null elements are truncated or hidden when not needed.

The MDM was built based on a graphical sketch of the powertrain system (figure 9.1.2) as agreed to by the team in a preliminary step. This schematic shows where physical connections exist between components documented in the components DSM. Additionally, mechanical, electrical, thermal, and chemical energy flows are shown by the use of colors and arrows. The directional and bidirectional nature of these flows serves as a guide for building the product functions DSM. For example, the Fuel Tank and the IC Engine components share a physical connection shown by the solid green arrow, which results in symmetrical edges above and below the diagonal between the two components in figure 9.1.1. The chemical energy of fuel, however, can only flow in one direction, which results in only one edge between the functions Store Fuel and Convert Fuel into Mechanical Energy above the diagonal in figure 9.1.1.

Two MDMs of the same size can be compared via matrix subtraction. The resulting MDM is labeled a ΔMDM (delta MDM) as presented in figure 9.1.3. The ΔMDM method can be used to compare two distinct architectures or two versions of a single architecture

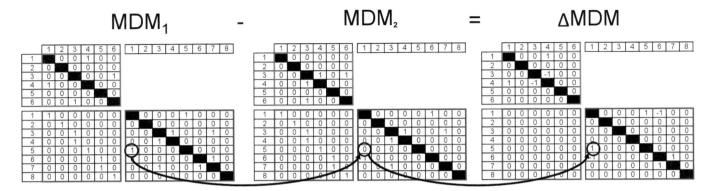

Subtraction by Fields

Figure 9.1.3
ΔMDM is computed as the difference between two MDMs of identical size.

that has been updated. Changes indicate that a component or function has been added or dropped.

Because of the binary nature of these MDMs, the ΔMDM results in matrix fields with values of {–1,0,1}. A ΔMDM matrix field value of {–1} shows a component or functional element present in architecture MDM_2 that is not contained in architecture MDM_1—as shown in figure 9.1.3. ΔMDM matrix field values of {0} denote no change, whereas a value of {1} indicates an interaction present in architecture MDM_1 not contained in architecture MDM_2.

ΣMDMs provide another useful analysis tool. The ΣMDM, referred to as a sum MDM or sigma MDM, is built by the addition of two or more MDMs as shown in figure 9.1.4. Similar to the ΔMDM, the matrices being added in a ΣMDM must match in terms of the function and component elements within the matrix position indexes.

Results

Two benefits of the ΔMDM method were readily recognized. First, the changes in components and functionality were easy to detect when comparing two architectures. Second, the method was useful in catching logical errors in matrices filled by hand in a workshop environment.

The addition of the eight MDMs within the set of vehicle concepts analyzed enabled the determination of which components apply to all architectures (cells showing a sum equal to the number of MDMs in the sum) and which components were found to vary across architectures. Architectures showing fields with a result of 1 indicate that the function or component is unique to one architecture from the original set. The information provided by the ΣMDM can be used to develop rules for design synthesis that specify

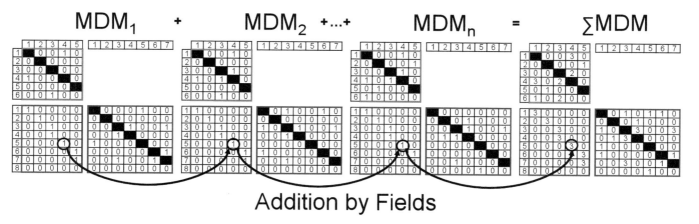

Figure 9.1.4
ΣMDM is computed by summing MDMs of identical size.

which component sets are necessary to perform a particular function or vice versa. It can also reveal components that are critical in all design variants.

The DMM portion of the ΣMDM is particularly useful. This DMM describes all connections between the component and functional domains across all architectures in the set. By turning the ΣDMM to a binary form, it can be used as a generic DMM in matrix manipulations when computing the function DSM given that a particular component DSM is known. This generic DMM thus enables DSM computations to explore function–component relationships of new structural configurations.

The ΣDMM is also useful in visualizing architecture information. For example, reading the DMM along a column shows the different components that map to the fulfillment of one function. Reading the DMM across rows displays the multiple functions a component can perform or that it is partly involved in performing.

In short, ΔMDMs and ΣMDMs offer a wide range of information when comparing product architectures. The methodologies presented can also apply to other types of architectures and MDMs.

References

Gorbea, Carlos. 2011, February. *Vehicle Architecture and Lifecycle Cost Analysis in a New Age of Architectural Competition*. PhD thesis, Technische Universität München (Institute for Product Development), München, Germany.

Gorbea, Carlos, Ernst Fricke, and Udo Lindemann. 2007, October 16–18. *Pre-Selection of Hybrid Electric Vehicle Architectures During the Initial Design Phase*. Proceedings of the 9th International DSM Conference, Munich, Germany, pp. 225–234.

Gorbea, Carlos, Tobias Spielmannleitner, Udo Lindemann, and Ernst Fricke. 2008, November 11–12. *Analysis of Hybrid Vehicle Architectures Using Multiple Domain Matrices*. Proceedings of the 10th International DSM conference, Stockholm, Sweden, pp. 375–387.

Example 9.2   Pratt & Whitney Jet Engine Product and Organizational Structures

Contributors

Manuel Sosa
INSEAD

Steven Eppinger
Massachusetts Institute of Technology

Craig Rowles
Pratt & Whitney

Problem Statement

Pratt & Whitney, a division of United Technologies Corporation, produces and supports aircraft jet engines, industrial gas turbines, and space propulsion systems. Development of a commercial aviation jet engine is a highly complex process, involving hundreds of engineers working simultaneously on the various components and subsystems. This two-domain DSM application investigated the system engineering and system integration aspects of the engine development process through the comparison of a product architecture DSM and an organization architecture DSM corresponding to the design of a commercial aircraft jet engine.

Data Collection

Over a period of four months in 1998, Craig Rowles (both an employee of Pratt & Whitney and a student in MIT's System Design and Management master's program) interviewed system architects in the PW4098 engine program (to capture the product architecture DSM) and lead engineers of the teams responsible for the design of all major physical and functional engine components (to capture the organization architecture DSM). Subsequent data codification, analysis, and interpretation of the DSM models were done jointly with Manuel Sosa, then a doctoral student at MIT. For a more thorough explanation of the product architecture DSM model, see example 3.2. For details of the organization architecture model, see example 5.3.

Model

The two-domain DSM model maps both the product and organization architectures of the PW4098 engine program by overlaying its (54 × 54 binary) organization DSM onto the corresponding (54 × 54 binary) product DSM, as shown in figure 9.2.1. This direct

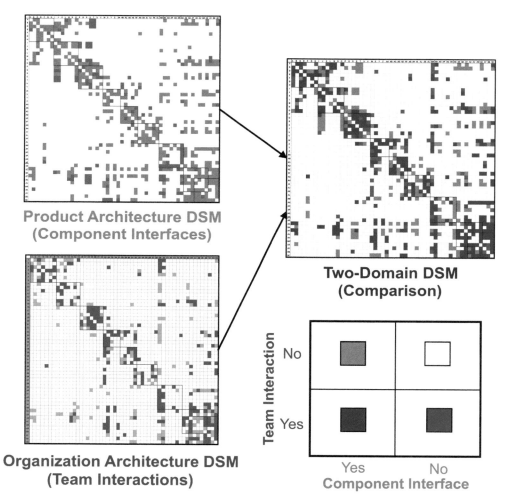

**Product Architecture DSM
(Component Interfaces)**

**Organization Architecture DSM
(Team Interactions)**

**Two-Domain DSM
(Comparison)**

Team Interaction

No

Yes

Yes   No

Component Interface

Figure 9.2.1
The product-domain DSM and the organization-domain DSM models were compared to identify areas of (mis) alignment between the product architecture and the organization architecture.

comparison of the DSM models across two domains is possible because there is one component design team in the organization DSM for each component in the product DSM. Both DSMs are sequenced identically, with clusters shown to represent boundaries of each subsystem (team). Each cell in the resulting alignment DSM corresponds to one of the following cases:

- **Matched team interaction and component interface** An interface between two components is matched by communication between the corresponding design teams (purple cells);

- **Matched lack of team interaction and component interface** No interface between two components corresponds to lack of communication between the corresponding design teams (blank cells);

- **Unmatched component interface** An identified interface between two components is not matched by technical communication between the corresponding design teams (red cells);

- **Unmatched team interaction** Two teams interact even though there is not an identified interface between the components designed by those teams (blue cells).

Results

The resulting two-domain comparison DSM not only captured the product–organization alignment during the design phase of the engine development but also the cases of product–organization misalignment. Although there was a significant alignment of the component interfaces and team interactions (almost 90% of the cells in the resultant DSM were either blank or purple), there was also a significant occurrence of misalignment (46% of the non-blank cells in the resultant DSM were either red or blue).

To investigate the misalignment, we studied several possible product and organizational factors that were systematically associated with the occurrence of mismatches of the component interfaces and team interactions. Three of the results of this analysis were:

1. We had collected information rating the criticality of each component interface in the product DSM. This allowed us to conduct an analysis to test the extent to which interface criticality matters. We found that less critical component interfaces were more often unmatched by team interactions.

2. We also had data regarding the nature of each of the component interfaces (spatial, materials, energy, etc.). Our analysis showed that some types of component interfaces were at higher risk of being unattended.

3. Because we knew which subsystem (team) was related to each component (team), we were able to analyze the interactions both within and across subsystems. This analysis

showed that mismatched interactions in both domains were more likely to occur across organizational boundaries corresponding to the subsystem definitions.

Our results helped engineering managers at Pratt & Whitney to better manage their complex system engineering challenges. Based on our analysis, they realized that a significant number of critical, unattended, and/or unidentified interfaces existed across subsystem boundaries. As a result, they applied more attention to identify and coordinate critical cross-boundary interactions.

References

Rowles, Craig M. 1999, February. *System Integration Analysis of a Large Commercial Aircraft Engine*. Master's thesis, Massachusetts Institute of Technology, Cambridge, MA.

Sosa, Manuel E. 2000, June. *Analyzing the Effects of Product Architecture on Technical Communication in Product Development Organizations*. PhD thesis, Massachusetts Institute of Technology, Cambridge, MA.

Sosa, Manuel E., Steven D. Eppinger, and Craig M. Rowles. 2004, December. The Misalignment of Product Architecture and Organizational Structure in Complex Product Development. *Management Science* 50 (12):1674–1689.

Sosa, Manuel E., Steven D. Eppinger, and Craig M. Rowles. 2007, November. Are Your Engineers Talking to One Another When They Should? *Harvard Business Review* 85 (11):133–142.

Example 9.3   Pitney Bowes Mailing System

Contributors

Anshuman Tripathy
Indian Institute of Management, Bangalore

Steven Eppinger
Massachusetts Institute of Technology

Problem Statement

Pitney Bowes is the world's largest vendor of mailing systems. Enabled by R&D innovation and advances in technology, and motivated by changes in postal regulations, Pitney Bowes products have evolved over the years from purely mechanical devices to complex mechatronic systems for processing mail at high speeds. This 1.5d DSM application explores the product development process of the MEGA mailing system in order to identify global product development opportunities for Pitney Bowes.

Data Collection

Through a series of interviews with engineers and managers at Pitney Bowes in 2006, Anshuman Tripathy (then a PhD student at MIT) documented the overall product development process for the MEGA mailing system (figure 9.3.1), including the product breakdown structure (PBS) of the system into components and their respective design and manufacturing organizations. We represented the development process using a process architecture DSM model and augmented this with information about the development dependencies between the product components and assignments to their respective design and manufacturing sources. The 1.5d DSM model of figure 9.3.2 was verified through discussions with Pitney Bowes personnel.

Model

The DSM shows the three phases of the development of the MEGA mailing system—system architecture, module development, and system integration. Decomposition of the MEGA mailing system, seen in the module development phase, identifies three modules—user interface, input, and finishing—each of which is comprised of several components represented in the DSM as engineering design tasks. The DSM shows three shaded groups of module development activities with coupling among the component development tasks within each module and little coupling across the three modules. The columns

Figure 9.3.1
A Pitney Bowes digital mailing system (courtesy of Pitney Bowes).

"Design" and "MfgEng/Prodn" identify whether Pitney Bowes or a supplier was responsible for the development and production of each component.

Results

The DSM shows that, following system architecture development, the three modules could each be developed quite independently. Such clean interfaces between modules were possible because Pitney Bowes spent a lot of effort (typically, approximately half of the product development project duration) in the system architecture phase of development. The DSM also shows that although each module has a primary manufacturing supplier producing many of its components, relatively little of the component design effort was conducted by these suppliers. We used the 1.5d DSM representation to provide Pitney Bowes new insight into the feasibility of further offshore product development of each of the modules.

At the time of our DSM analysis of the MEGA mailing system, the user interface module was being manufactured primarily by a single supplier, Cherry. The DSM suggests that most of the development work of this module could eventually also be outsourced to the same supplier. However, the PSD and software/chip would need to be controlled closely due to security considerations, so these would likely not be assigned to Cherry.

Most of the input module components were being manufactured and assembled by Brother. This supplier was also known for its engineering capabilities and could be considered for the complete design, development, production, and testing of the input module. This could include the power supply unit, which was then being developed and supplied by various suppliers, and comprised of standard parts.

Finally, the design and development of the entire finishing module, with the exception of the MMC (motion controller), which was considered a core technology, could feasibly

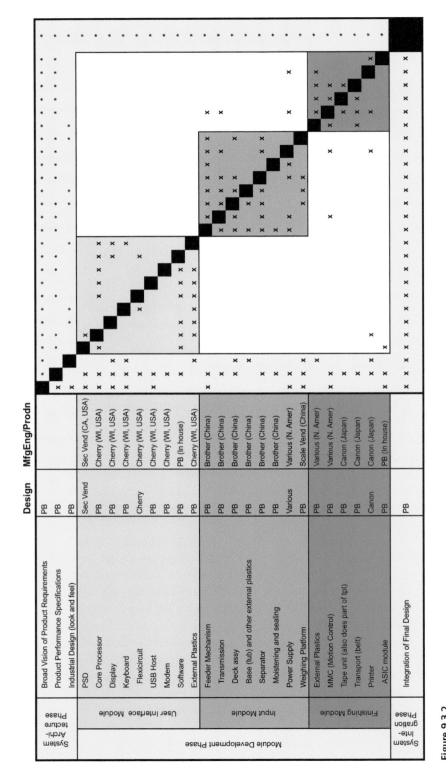

Figure 9.3.2
System architecture DSM for Pitney Bowes MEGA mailing system.

be outsourced to Canon. Canon, in Japan, was already developing and supplying the printer unit.

The coupling within each of the module development activity blocks would appear to favor outsourcing all of each module's component design effort to a single supplier. This would also facilitate the integration of each module. Furthermore, given the presence of such clean interfaces between the modules, Pitney Bowes would likely be able to manage the final system integration effectively. Our analysis was presented to Pitney Bowes managers who were then able to develop their design outsourcing strategy accordingly.

References

Tripathy, Anshuman. 2010, January. *Work Distribution in Global Product Development Organizations*. PhD thesis, Massachusetts Institute of Technology, Cambridge, MA.

Tripathy, Anshuman, and Steven D. Eppinger. 2011, August. Organizing Global Product Development for Complex Engineered Systems. *IEEE Transactions on Engineering Management* 58(3):510–529.

Example 9.4   Audi AG Team Composition for Collaboration

Contributors

Matthias Kreimeyer
Institute of Product Development, TUM

Ulrich Herfeld
Audi AG

Problem Statement

To enable efficient collaboration among different disciplines, cross-functional team structures are an important enabler. This example shows how DMMs were used to help formalize interactions between design and simulation departments at Audi AG, a major premium automobile manufacturer in Germany. This engineering process involves an organization of approximately 800 engineers in embodiment design and about 50 engineers in simulation, who were part of the development of the so-called "trimmed body" of a sedan. This scope comprises the car's body, all doors and hatches, as well as the interior paneling, including about 400 components exposed to approximately 130 numerically simulated load cases related to comfort and safety.

Within the design process, individual engineers need to collaborate with a multitude of colleagues to cooperatively design a highly integrated product. This DMM application was therefore aimed to support these engineers with a structure to communicate with other engineers as needed to establish different functions (represented as load cases) to validate the vehicle's components and their interfaces. We focused our analysis to answer the question of how, through teams of manageable size, coordination of all engineers could be achieved so that, at the same time, information transfer in both directions could be ensured.

Data Collection

The scope of our data collection included all engineering functions concerned with the trimmed-body finite-element simulation for NVH load cases only (noise, vibration, and harshness requirements). Three weighted DMMs were built:

1. responsibility of embodiment design engineers for components
2. involvement of components in simulated load cases
3. responsibility of simulation engineers for load cases to be simulated

The DMM models were built as weighted matrices to represent the degree of involvement of one domain in the next. The data were collected mostly from documentation of functional and technical specifications for the overall vehicle project. For each component or subsystem, specific load cases and simulations are necessary to receive technical approval. These data were collected through interviews of the simulation engineers.

The component responsibility data were obtained from the specification documents that contained a basic work breakdown structure (WBS). This attribution of responsibilities was then refined based on Audi's phone book and interviews in both simulation and design departments.

Model

In each DMM, the strength of the involvement between domains was expressed as a score using a 3-point scale. For the second DMM, mapping components to load cases, the following scale was used:

- Level 3—component is evaluated by load case (strongest linkage)
- Level 2—component is a significant part of the model
- Level 1—component is an element of the model's border area

Depending on how much an engineer is responsible for a component (in embodiment design) or a load case (in simulation), his or her involvement in that element is scored accordingly in the DMM. This ensures at a later stage that those people of higher relevance to a cluster of components and load cases can be identified. In the first and third DMMs, the following scale was used:

- Level 3—engineer is responsible for component or load case (strongest linkage)
- Level 2—engineer conducts the embodiment design/simulation of component/load case
- Level 1—engineer supports embodiment design/simulation of component/load case

We found that component and simulation responsibilities were often not formalized. Ultimately, therefore, only levels 2 and 3 were used because level 1 was too fuzzy to serve as a basis for consistent data acquisition.

Figure 9.4.1 shows the DMM mapping components to load cases.

Results

To obtain teams, in a first step, the component to load case DMM was clustered, as shown in figure 9.4.1. Each cluster contains a set of components and load cases that are similar

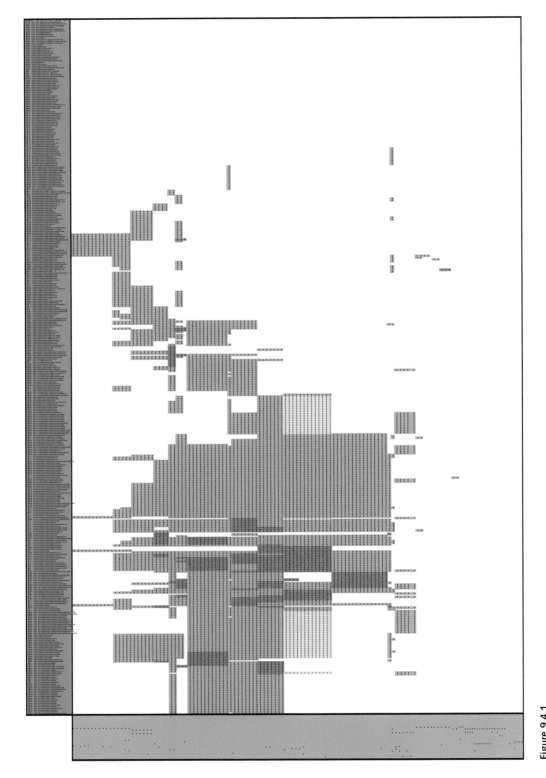

Figure 9.4.1
Clustered DMM mapping components (in columns) to load cases (in rows) with weights (3 = red, 2 = orange, 1 = yellow).

from the structural point of view, as the components serve the load cases in a comparable manner and vice versa. A cluster is therefore a building block across which communication between design and simulation engineers can be aligned. However, the designation of clusters needs to be done carefully to obtain good clusters (i.e., where both components among each other and load cases among each other are comparable in the way one domain serves the other). The clustering was, therefore, done manually involving much discussion with the engineers over two days to ensure proper interpretation of the results. We found that the difficulty was not in finding the clusters but in breaking larger clusters into smaller ones.

We joined the three DMMs as illustrated in figure 9.4.2. By doing so, we could determine which engineers were involved in each cluster in the component-to-load case DMM. Depending on the size of the initial cluster, these teams could be large. In such a case, one large team is not desirable, so clusters were decomposed to a more manageable size.

To support the clustering decisions, the level of interaction between any two organizational units can be calculated as follows: If an engineer is only supporting the embodiment design (weight 1) of one component that only borders the simulation area (weight 1) and only has to interact with a simulation engineer who is supporting the simulation of a load case (weight 1), then their interaction strength is low ($1 * 1 * 1 = 1$). If, however, another person is responsible for the embodiment design of a component (weight 3) and conducts the embodiment design of yet another component (weight 2), and each component is evaluated (weight 3) by a simulation engineer responsible for that load case (weight 3), then the interaction strength comes to $3 * 3 * (3 + 2) = 45$.

Each of these cluster-based teams described a set of components that related to a set of load cases. Therefore, these clusters needed to be collected to evaluate load cases or components. The clusters served as building blocks and were combined to form teams. Some teams integrated all design engineers involved in a cluster of load cases. Other teams involved all simulation results relevant to a cluster of components, forming a functional integration team. A total of 153 clusters were combined into 12 teams for function evaluation and 22 teams for the integration of functions into components.

To generate a core team that could supervise the overall activities of these 34 teams, only interactions at level 3 were considered for all three matrices. In doing so, the component-to-load cases DMM in figure 9.4.2 contains only rows and columns with at least one red element. The DMM for the simulation engineers was rather small because only six simulation departments were involved in the 65 remaining load cases; 32 out of 153 clusters were identified as core clusters with relevant level 3 relationships that contribute to the coordination team. Figure 9.4.2 also demonstrates how team building blocks were constituted from the clusters of the component-to-load case DMM.

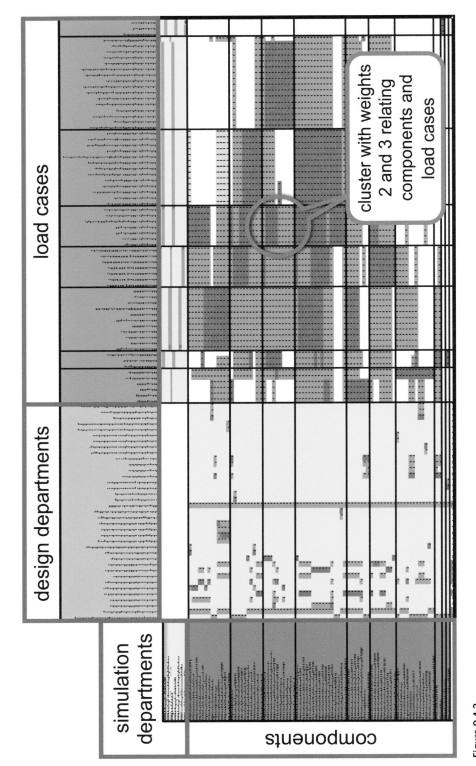

Figure 9.4.2
The three DMMs, with the clustered components to load cases DMM (lower right) used to deduce team building blocks, showing only elements with one or more level 3 weights (red cells).

References

Herfeld, Ulrich, Matthias Kreimeyer, Frank Deubzer, Tobias Frank, Udo Lindemann, and Ulrich Knaust. 2006. *Verknüpfung von Komponenten und Funktionen zur Integration von Konstruktion und Simulation in der Karosserieentwicklung. Berechnung und Simulation im Fahrzeugbau*, pp. 259–276. VDI Wissensforum IWB GmbH. (in German)

Kreimeyer, Matthias, Frank Deubzer, Mike Danilovic, Stefan Daniel Fuchs, Ulrich Herfeld, and Udo Lindemann. August, 2007. *Team Composition to Enhance Collaboration between Embodiment Design and Simulation Departments*. Proceedings of the International Conference on Engineering Design, ICED'07, The Design Society, Paris, France.

Example 9.5   U.S. Senate

Contributor

Jason E. Bartolomei
Massachusetts Institute of Technology

Problem Statement

For many, the U.S. Senate is a complex organization that is difficult to understand. Actually, the senate is both structurally and behaviorally complex. In this example, the MDM and DSM are used to understand the relationships between senate offices through each senator's committee assignments. As many know, the control over the legislative process happens in committee. In the senate, members are assigned to several committees. The relationships that senators share through their committee assignments provide an important component to understanding the structure of the Senate. By better understanding this structure, Senate staffs, committee staffs, and external organizations are better able to develop engagement strategies for legislation. This example uses the MDM as

Figure 9.5.1
U.S. Capitol building (courtesy of Architect of the Capitol).

a means to visualize and analytically examine the structure of the U.S. Senate (111th Congress).

Data Collection

Every year a number of organizations publish the Senate Committee Assignments. This publicly available information was used to construct the MDM.

Model

The MDM consists of two matrices. The first, a DMM (see excerpt in figure 9.5.2) relates each senator's office (SO) to its corresponding subcommittee assignments (CTE).

Using the information provided in this DMM, we are able to determine the connections between SOs through their committee assignments by simply squaring the matrix:

$$[CTE \times SO]^T [CTE \times SO]$$

This yields the SO DSM [SO × SO] shown in figure 9.5.3.

Results

The SO DSM shown in figure 9.5.3 provides a means for understanding the structure of the Senate. The diagonal cells (i,i) indicate the number of committee assignments for each senator, with green shading for those with fewer than 10 committees and red shading for 10 or more committees. Each off-diagonal cell (i,j) represents the sum of the

	Akaka	Alexander	Barrasso	Baucus	Bayh	Begich	Bennet	Bennett	Bingaman	Bond	Boxer	Brown (MA)	Brown (OH)	Brownback	Bunning	Burr	Burris	Byrd	Cantwell	Cardin	Carper	Casey	Chambliss	Coburn	Cochran	Collins	Conrad	Corker	Cornyn	Crapo	Demint	Dodd	Dorgan
Senate Committee on Health, Education, Labor, and Pensions																																	
* Subcommittee on Retirement and Aging	1				1		1									1						1		1									
* Subcommittee on Children and Families	1						1							1								1		1								1	
* Subcommittee on Employment and Workplace Safety							1							1		1																1	
Senate Committee on Indian Affairs	1	1															1							1				1		1			1
Senate Select Committee on Intelligence					1					1							1						1	1									
Senate Committee on the Judiciary																																	
* Subcommittee on Administrative Oversight and the Courts																				1													
* Subcommittee on Antitrust, Competition Policy and Consumer Rights																														1			
* Subcommittee on The Constitution																				1				1						1			
* Subcommittee on Crime and Drugs																				1				1									
* Subcommittee on Immigration, Refugees and Border Security																													1				
* Subcommittee on Human Rights and the Law																				1				1						1			
* Subcommittee on Terrorism and Homeland Security																				1				1						1			

Figure 9.5.2
Excerpt of the SO to CTE DMM.

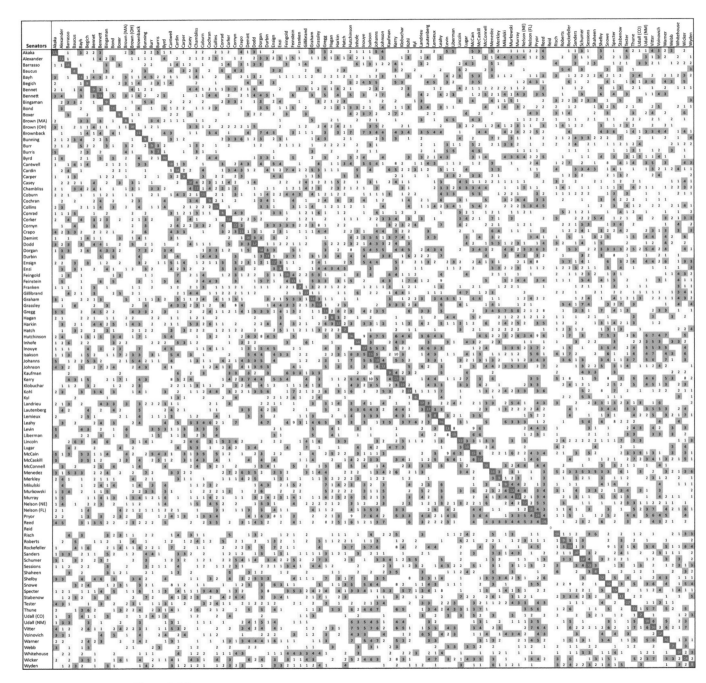

Figure 9.5.3
DSM showing relationships between senate offices.

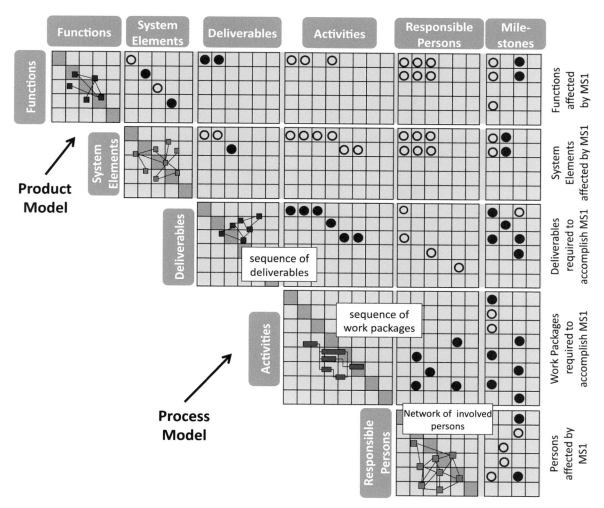

Figure 9.6.2
Layout of the MDM used to analyze the product and its development process.

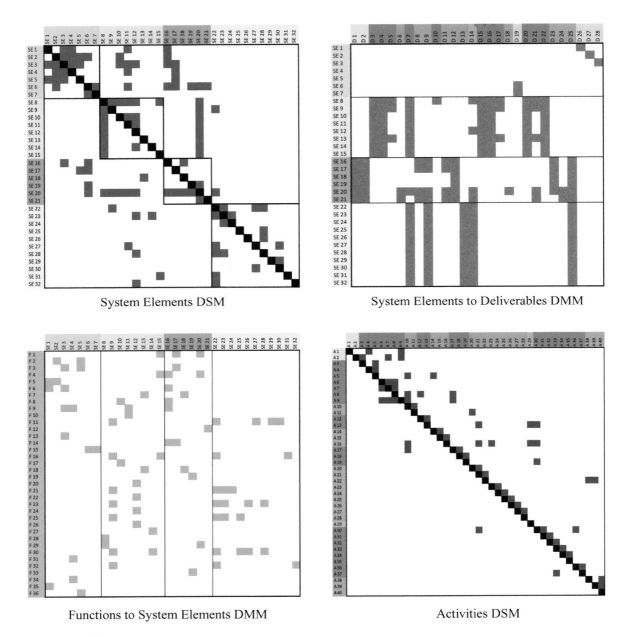

System Elements DSM

System Elements to Deliverables DMM

Functions to System Elements DMM

Activities DSM

Figure 9.6.3
Two DSM and two DMM excerpts from the MDM model.

diagonal DSMs and their adjacent DMMs) (Lindemann et al. 2009). The computation of indirect interdependencies based on the known interconnections can be done (e.g., by using a single DMM to derive two DSMs or by combining a DSM of one domain and a neighboring DMM to compute a DSM in another domain). Altogether there are six computational logics. The identification and analysis of indirect dependencies is of special interest because our interviews showed that these unknown, indirect dependencies are responsible for most of the problems and delay within the process.

Results

The integrated product and process model of the sunroof offered a high variety of possibilities to analyze and optimize the system. Some examples are described below.

Within the product model (functions, system elements), it was possible to deduce the functional structure from (shared) system elements. The graphical representation of this structure enabled the engineers at BMW to realize and understand hidden dependencies among the components and functions even across disciplines and departments. These indirect dependencies, especially across departments, were often surprising and responsible for unsuccessful tests.

In the case of design changes, the model enabled the developers to notice which functions or other components were affected. This is particularly important because the functional responsibility is often located in a different department. As a result, they were able to see who had to be informed of the change and which functions had to be checked. In addition, the linkage of product and process models also offered the possibility to easily trace the impacts on the process structure of changes in the product structure.

The combination of the derived functional structure with the DMM mapping functions to responsibilities allowed the deduction of a network of responsibilities based on common functions. Analysis of the responsibilities DSM showed that there were two kinds of responsibilities involved within the project. On the one hand, a highly interconnected core team is responsible for the technical development of the sunroof. On the other hand, a lot of departments are only involved in certain aspects such as testing or the vehicle interior. This information can be used for the composition of teams or the optimization of BMW's organization structure.

The model also allowed associating the project milestones with functions and system elements through the deliverables, which meets well-established working routines in departments of different disciplines. The engineers were able to see immediately which milestone affects which function and system element and could use that information for their personal work planning.

Another issue is planning future development processes. Using the MDM model makes it possible to derive a simplified sequence of the work packages. Assuming the basic logical structuring of the process by milestones where customer functions are tested, the time between two milestones is determined by the work packages necessary to produce the deliverables within that development phase. Starting from a milestone, the work packages can be arranged forward or backward in time, where the sequence of work packages can be deduced from the sequence of deliverables.

References

Braun, Stefanie Constanze, Holger Diehl, Markus Petermann, David Hellenbrand, and Udo Lindemann. 2007, October. *Function Driven Process Design for the Development of Mechatronic Systems*. Proceedings of the 9th International DSM Conference, Munich, Germany.

Fischer, Markus, Stefanie Braun, David Hellenbrand, Christian Richter, Olaf Sabbah, Christian Scharfenberger, Michael Strolz, Patrick Kuhl, and Georg Färber. 2008, September. *Multidisciplinary Development of New Door and Seat Concepts as Part of an Ergonomic Ingress/Egress Support System*. Proceedings of the FISITA World Automotive Congress, FISITA (UK) Limited, Munich, Germany.

Lindemann, Udo, Maik Maurer, and Thomas Braun. 2009. *Structural Complexity Management: An Approach for the Field of Product Design*. Berlin: Springer.

Example 9.7   HILTI Adhesive Anchors and Dispensers

Contributors

Maik Maurer and Alexander Suessmann
Technische Universität München

Andreas Schell
HILTI Entwicklungsgesellschaft mbH

Problem Statement

HILTI Entwicklungsgesellschaft is a division of the HILTI Group, a provider of tools, systems, and services for the global construction industry. HILTI's HIT system is a two-component, adhesive injection system (figure 9.7.1) for heavy duty anchoring in concrete. The system consists of different dispensers and foil cartridges containing the adhesives. Ongoing development of the HIT system includes many product improvements over time. Due to the complexity of the integrated system, even small changes to some components can require significant changes in others. Moreover, it proved difficult for experts of different areas such as foil design, mortar development, and mechanical engineering to take all interdisciplinary side effects into account. This MDM application was intended to help ensure successful product development with the following objectives:

Figure 9.7.1
HILTI's HIT adhesive injection anchoring system (courtesy of HILTI Entwicklungsgesellschaft).

1. Visualize developers' interdisciplinary change impact.

2. Identify suitable possibilities for adjusting the system's configuration and support the design of experiments (DoE).

3. Identify measures for better control of the system's complexity.

Data Collection

Over a period of four months in 2010, Alexander Suessmann, a graduate student at the Technische Universität München, conducted interviews with experienced developers of the HIT system at HILTI Entwicklungsgesellschaft. After decomposing the product system into components, he inquired about: (1) design parameters that can be directly influenced, and (2) indirect relational characteristics of the entire system, which are determined by the design parameters. A certain shape of a component, for instance, represents a design parameter; this determines the flow resistance, which is a relational characteristic.

Prior to the collection of dependencies, the design parameters were classified into domains of shape, material, type/state, production parameter, environment parameter, and mortar ingredients in order to define distinct types of direct interactions (e.g., geometrical).

The acquisition of dependencies in the MDM was executed subset by subset. Within a half-day workshop, typically one or two subsets could be completed. An important advantage of the decomposition of the entire system into subsets was that for every workshop, only the required experts, typically two or three, had to participate.

Model

Figure 9.7.2 shows the two main domains represented in the MDM—design parameters (DPs) and relational characteristics (RCs). The meaning of the dependencies is noted in

	Design Parameters	Relational Characteristics
Design Parameters	can change	determine
Relational Characteristics	✕	influence

Figure 9.7.2
Basic MDM layout for the HIT system.

the three matrices. The domain in a row is mapped to the domain in a column (IC/FBD convention). For example, design parameters can change other design parameters, whereas design parameters can determine relational characteristics. The lower left part of the MDM does not contain problem-relevant dependencies and can be excluded.

The MDM shown in figure 9.7.3 decomposes both main domains into subgroups. The RC domain additionally contains a subgroup named System. It relates to RCs that cannot be allocated to a single component. For example, the final mixing quality and mortar volume per stroke are typical system RCs. The total number of elements is about 300. Only dark gray-shaded matrices contain dependency information; 110 of the 169 possible matrices were able to be excluded from further data acquisition, which meant that not all possible dependencies in the entire MDM had to be considered. Therefore, the system modeling process could be executed efficiently.

Figure 9.7.4 shows details of the portion of the DMM marked *** in figure 9.7.3. It shows exemplar DPs and RCs of the foil packs. Cells containing a 1 indicate that a DP (row) determines a RC (column), whereas a 0 indicates that the dependency has been discussed and no influence has been identified.

Results

The MDM layout depicted in figure 9.7.3 gives an outline of the system as it summarizes the component dependencies. For example, the Foil Pack DPs influence not only the RCs of the Foil Pack but also those of the Foil Cartridge and the System. For investigating details such as which DPs can change which DPs and influence which RCs, we examined the particular DSMs and DMMs, respectively.

Alternatively, the MDM can be analyzed at the element level of detail. For these investigations, we also utilized graph theory. For instance, the force-directed graph shown in figure 9.7.5 illustrates the entire MDM with elements indicated by IDs. We colored the DPs and RCs to show each component's contribution to the system's RCs (pink) and their mutual influence. Elements and dependencies on the edge of component clusters represent their interfaces to other components. They can be itemized by focusing on the active and/or passive surrounding of single elements. Altering elements on the edge of a component cluster can necessitate changes to other components. This is illustrated in the right side of figure 9.7.6, where the direct surrounding of a connector DP (green) is shown. The DP determines RCs of the system (pink) and the foil packs (orange). Moreover, its modification potentially changes several other DPs allocated to the dispenser (blue) and the mixer (red). Those changes often remain unrecognized as they concern interdisciplinary responsibility. By means of feedforward analyses conducted on those elements (see left side of figure 9.7.6), we were able to systematically rule out and discuss all potential change propagations (e.g., the outlet geometry of the connector indirectly influences the mortar's mixing quality).

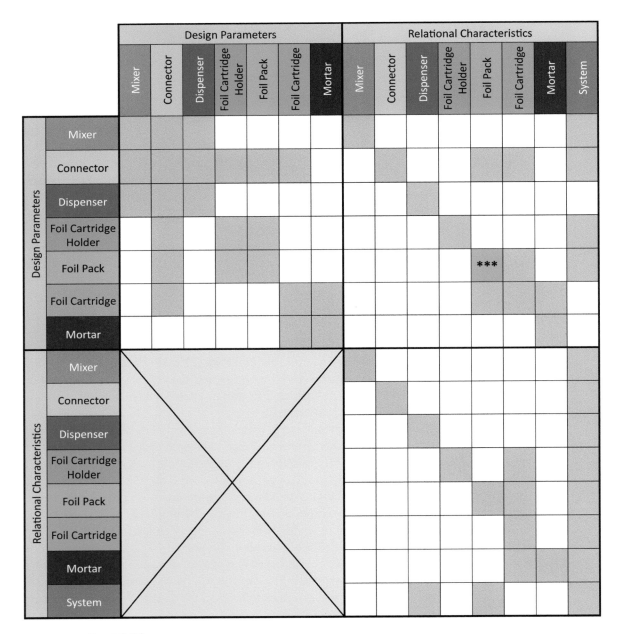

Figure 9.7.3
More detailed MDM layout of the HIT system, with shaded subsets containing direct dependencies.

	RC FP Breaking strength A	RC FP Breaking strength B	RC FP Piercing strength A	RC FP Piercing strength B	RC FP Volume A	RC FP Volume B	RC FP Clip stability A	RC FP Clip stability B	...
DP FP Thickness of layers A	1	0	1	0	0	0	0	0	
DP FP Thickness of layers B	0	1	0	1	0	0	0	0	
DP FP Clip type A	0	0	0	0	0	0	1	0	
DP FP Clip type B	0	0	0	0	0	0	0	1	
DP FP Length A	0	0	0	0	1	0	0	0	
DP FP Length B	0	0	0	0	0	1	0	0	
DP FP Diameter A	0	0	0	0	1	0	0	0	
DP FP Diameter B	0	0	0	0	0	1	0	0	
...									

Figure 9.7.4
Excerpt regarding foil packs from the DP-to-RC DMM.

The DP and RC mindset used in the local MDM definition also complies with the design of experiments used in the development process of the HIT system. DPs and RCs correspond to parameters to be varied and characteristics to be measured, respectively. Therefore, we used the active and/or passive surrounding also for navigating through the structure and for systematically identifying and evaluating suitable DPs for determining certain RCs. The passive surrounding of an RC reveals all DPs determining it and all RCs having an influence on it. If no suitable DPs were in the direct surrounding of a particular RC, then we used one of the ambient RCs for determining it and regarded its passive surrounding. Once a potentially suitable DP was detected, it was evaluated by a feedforward analysis that revealed all the side effects of altering it (e.g., the connection of outlet geometry and mortar mixing quality as mentioned earlier).

We also applied a structural Pareto analysis to the RC DSM to identify measures for better control of the system's complexity. The analysis disclosed that an RC of the dispenser contributed to about 2,000 potential feedback loops allocated to three incoming relations. RCs like this should be stabilized or designed more robustly. We weakened the impact of relevant RCs to simplify the system's configuration. Therefore, we identified

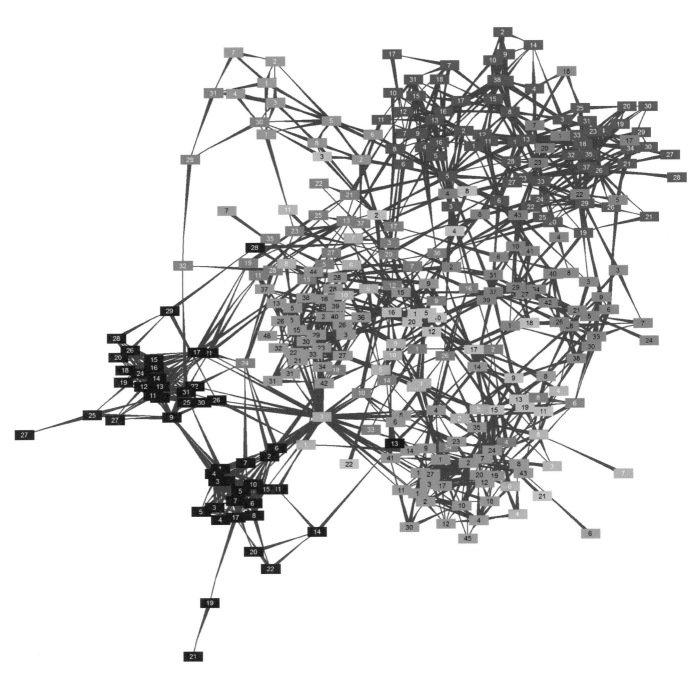

Figure 9.7.5
Force-directed graph representation of the entire MDM.

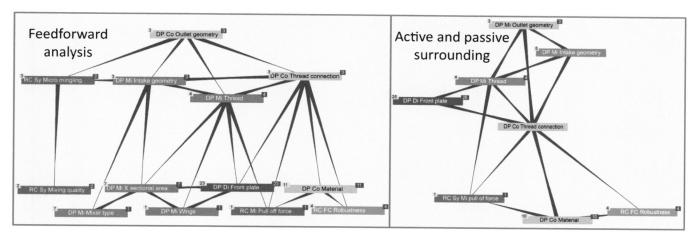

Figure 9.7.6
Details of a feedforward analysis (left) and a direct surrounding view (right) of two connector DPs in figure 9.7.5.

adequate DPs (as described earlier) and consequently increased the robustness of the significant RC. Thus, the system became easier to handle.

References

Lindemann, Udo, Maik Maurer, and Thomas Braun. 2009. *Structural Complexity Management: An Approach for the Field of Product Design.* Heidelberg: Springer.

Suessmann, Alexander. 2010, October. *Structural Complexity Management in the Field of Adhesive Anchor System Design.* Diploma thesis, Technische Universität München, Munich, Germany.

Example 9.8   Digital Research Labs Change Packaging in Systems Design

Contributors

Naveed Ahmad, David Wynn, and John Clarkson
Engineering Design Centre, University of Cambridge

Problem Statement

Engineering changes are ubiquitous in development projects. They are unpredictable in terms of when they occur and also in terms of their nature and how they propagate. In practice, changes that arise unexpectedly during a project are often put aside and allowed to accumulate because executing them together can reduce overheads such as task setup time and testing after performing the rework. However, it can also create additional work while changes wait to be processed because tasks are done that may have to be revisited when the changes are eventually processed. This raises interesting questions when deciding how to process changes and an opportunity for an MDM-based simulation model to assist in identifying the most appropriate change processing approach for a given project.

Data Collection

A simulation model was constructed for a microcontroller-based device, a product of Digital Research Labs, an engineering design company based in Pakistan. The model was constructed in three stages by eliciting a product subsystem DSM, a process DSM, and a product-to-process DMM in which each subsystem is connected to at least one activity that contributes to its definition in the detailed design process. In this context, such links are assumed to be directional. The design was modeled by the authors using documentation provided by the manufacturer and supplemented by telephone interviews with the designers. The model was then sent by e-mail to the designers, who verified that it represented their product and process. The procedure for building the product DSM consisted of decomposing the product into subsystems/components, recording the linkages between these items, and estimating the impact and likelihood of a change propagating directly between each pair of subsystems. Likewise, the procedure for building the process DSM consisted of decomposing the product development process into activities and identifying the associated input and output information. The total time required to build the product and process DSMs, as well as the linking DMM, was about eight hours.

Model

The MDM has the product DSM in the upper left corner and the larger process DSM in the lower right corner (see figure 9.8.1). The product DSM takes the form of a change

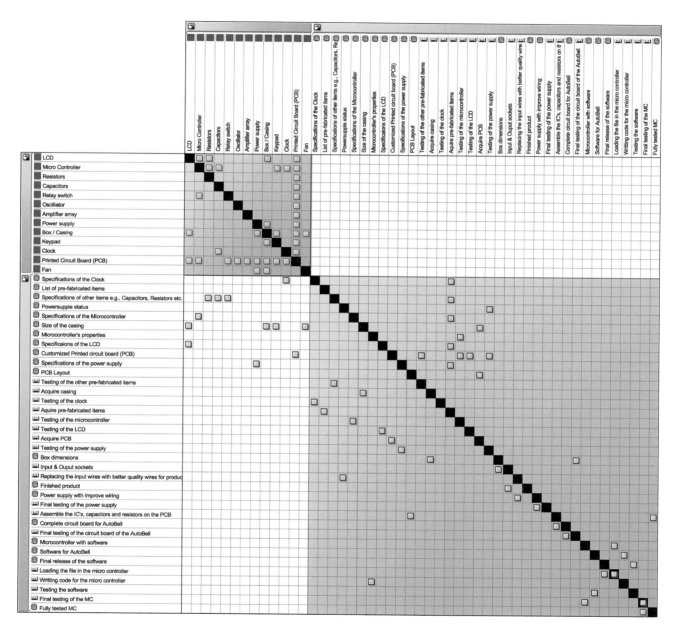

Figure 9.8.1
MDM for microcontroller product and design process.

propagation DSM with marks at the intersections representing the risk of change propagating between subsystems (see example 3.6 for an explanation and application of the change propagation DSM). The process DSM and the product-to-process DMM are binary, with marks of uniform size representing a link. The combined MDM was used as the basis for a simulation experiment to identify the tradeoff between change review interval (the time between processing changes) and the average delay to project completion. In the experiment, each change request is processed in three steps: (1) the product DSM is used to identify the components, (2) the product-to-process DMM is used to identify the activities directly related to these components, and (3) the process DSM is used to identify all the activities requiring rework. These activities are then stored in a buffer for a fixed interval after which their execution is simulated.

Results

Early results indicated that an optimal review interval to limit project delay could be found (figure 9.8.2) assuming a realistic sequence of changes; in this case, a delay of a little more than 30% of the duration is expected if no changes occur. Changes occurring during a design project may be executed immediately or left to accumulate in batches. The modeling and analysis performed in this study highlighted the need to choose an appropriate change processing interval to minimize the overhead of unnecessary rework. The results were of great interest to Digital Research Labs, and investigations are ongoing to explore the merits of changes to their processes.

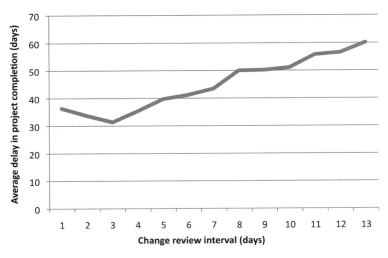

Figure 9.8.2
Average delay in project completion for varying intervals.

This model can be developed further to incorporate other factors influencing the execution of changes—most important, the availability of resources to execute change requests. The simulations also highlight the number of times each activity was reworked and the amount of rework in different activities, data that could be utilized to adjust the product or process architecture to make projects less sensitive to repeated rework.

References

Ahmad, Naveed, David Wynn, and John Clarkson. 2010, July 22–23. *The Impact of Packaging Interdependent Change Requests on Project Lead Time*. Proceedings of the 12th International Dependency and Structure Modeling Conference, Cambridge, England.

Ahmad, Naveed, David Wynn, and John Clarkson. 2010, October 20–22. *When Should Design Changes Be Allowed to Accumulate?* Proceedings of IDMME—Virtual Concept, Bordeaux, France.

Clarkson, John, Caroline Simons, and Claudia Eckert. 2004. Predicting Change Propagation in Complex Design. *Journal of Mechanical Design* 126 (5):765–797.

Example 9.9   Airport Security System

Contributors

Maik Maurer
Teseon GmbH

Mara Cole
Bauhaus Luftfahrt e.V.

Problem Statement

Civil aviation faces a constant threat from terrorist attacks. The airport functions as a gateway, and installed security checkpoints are meant to reduce the occurrence of attacks. Being able to cope in an efficient way with both potential threats and increasing passenger volume is a highly demanding challenge. To prepare the airport for future threats, one needs to take a systems view in order to thoroughly understand the elements of possible future threat scenarios as well as their interrelation with existing security measures.

Data Collection

Bauhaus Luftfahrt is an international think tank founded by the Bavarian Ministry for Economic Affairs and three aerospace companies, EADS, Liebherr-Aerospace, and MTU. Together with Teseon, a software development and consulting company, Bauhaus Luftfahrt constructed an airport security system MDM model containing approximately 300 elements grouped into 15 domains. Within this system, there are approximately 11,000 possible relations, of which more than 3,200 direct dependencies were specified. At first, we identified the relevant elements in brainstorming sessions with up to six experts and a moderator. The identified elements were directly depicted in a mind map and then classified in a hierarchical tree structure. Elements describing the main branches of this structure served as the 15 domains for the MDM model, structured as shown in figure 9.9.1.

The 15 domains in the square MDM resulted in 225 submatrices describing general dependencies within and between the domains. In a subsequent step, relevant submatrices with direct dependencies were identified and characterized. For example, the domain *tool/weapon* is linked directly to the domain *use of tool/weapon* (by the relation *allows*) but not to the domain *intention of offender*. It turned out that fewer than 20% of the submatrices were directly dependent and consequently utilized for the system modeling.

Finally, we transferred the system elements from the mind map to the MDM as row and column elements in their respective domains. In a series of workshops, the element

	use case		threat scenario								airport layout				
	actor	use	potential offender	intention of offender	tool/weapon	use of tool/weapon	approach of offender	insertion of tool/weapon	target	threat	departure zone	end zone	attack zone	security activity	security technology
actor		can carry out												can lead to	
use			excludes		excludes	excludes	excludes	excludes	excludes	excludes	excludes sojourn	excludes sojourn		induces	
potential offender				has						allows		is located in			
intention of offender			correlates with												
tool/weapon						allows		suitable for	reachable through	allows					
use of tool/weapon									suitable for						
approach of offender								allows	suitable for	allows			suitable for		
insertion of tool/weapon										allows	leads to				
target									correlates with	can lead to			is located in		
threat															
departure zone												has follower			
end zone															
attack zone												is situated in			
security activity							can impede	can counteract			is situated in	is situated in	is situated in		can apply
security technology					can detect										

Figure 9.9.1
Layout of the MDM for describing valid threat scenarios.

		target													
		aircraft - in flight	aircraft - on ground	airport - outside control tower	airport - inside control tower	airport - fuel depot	airport - apron	airport - maintenance facilities	people - restricted area	people - security area	people - public area	airport - freight terminal	communication - ground-air	information technology	airport - infrastructure
intention of offender	economic loss	1	1	1	1	1	1	1	0	0	0	1	1	1	1
	human life	1	1	0	0	0	0	0	1	1	1	0	0	0	0
	attention, headlines	1	1	1	0	1	1	0	1	1	1	0	0	0	1
	fear, demoralisation	1	1	1	0	1	1	0	1	1	1	0	0	0	1
	survival, escape	1	1	0	0	0	0	0	0	0	0	0	0	0	0
	base motives - personal gain	1	1	0	0	0	0	0	0	0	0	0	1	0	0
	base motives - murder	1	0	0	0	0	0	0	1	1	1	0	0	0	0
	mentally disturbed	1	1	1	1	1	1	1	1	1	1	1	1	1	1
	blackmail	1	1	1	1	1	1	1	1	1	1	1	1	1	1
	none	0	0	0	0	0	0	0	0	0	0	0	0	0	0

Figure 9.9.2
DMM showing direct dependencies between the *intention of offender* and *target* domains.

dependencies indicated by the direct interrelation of the respective domains were specified. See figure 9.9.2 for an example DMM.

Model

The identified domains can be aligned by triangularization, resulting in a clear sequence for the composition of valid threat scenarios as illustrated in figure 9.9.3. Starting the scenario-building process, the first two domains indicate a person's apparent use of the airport infrastructure. Whether somebody goes shopping or on an international flight affects which kind of security measures he might be confronted with and which areas of the airport he might have access to. This definition already narrows down the element choice for the subsequent scenario generation (figure 9.9.3, group 1). For example, somebody shopping at the airport will not be able to reach the target, *aircraft—on ground*, because he will not be granted access to secure areas.

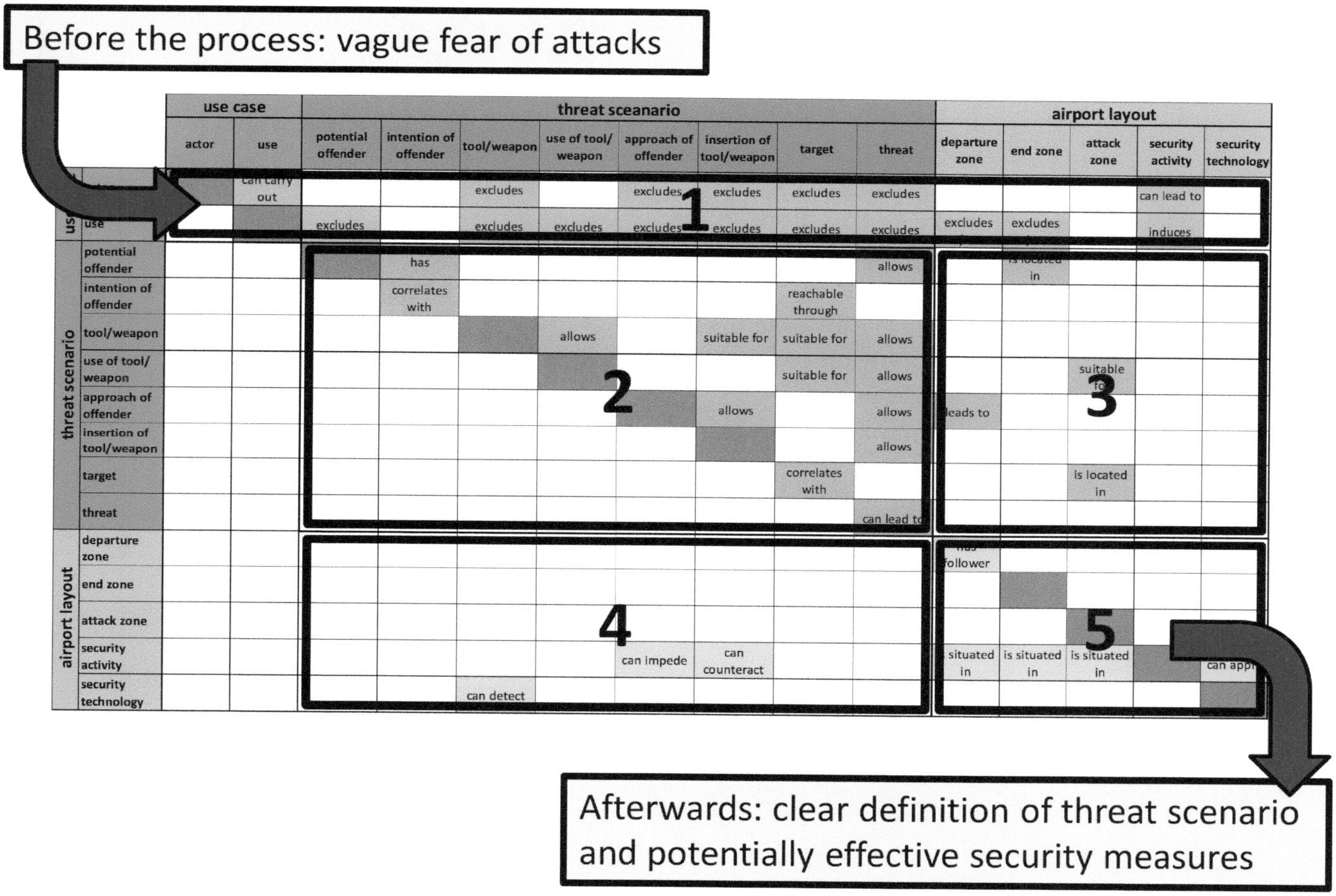

Before the process: vague fear of attacks

		use case		threat sceanario								airport layout				
		actor	use	potential offender	intention of offender	tool/weapon	use of tool/ weapon	approach of offender	insertion of tool/weapon	target	threat	departure zone	end zone	attack zone	security activity	security technology
use case	can carry out					excludes		excludes	excludes	excludes	excludes				can lead to	
	use			excludes		excludes	excludes	excludes	excludes	excludes	excludes	excludes	excludes		induces	
threat scenario	potential offender				has						allows		is located in			
	intention of offender				correlates with					reachable through						
	tool/weapon						allows		suitable for	suitable for	allows					
	use of tool/ weapon									suitable for	allows			suitable to		
	approach of offender								allows		allows	leads to				
	insertion of tool/weapon										allows					
	target									correlates with				is located in		
	threat										can lead to					
airport layout	departure zone												has follower			
	end zone															
	attack zone															
	security activity							can impede	can counteract			situated in	is situated in	is situated in		
	security technology					can detect										can apply

Afterwards: clear definition of threat scenario and potentially effective security measures

Figure 9.9.3
MDM structured into groups of DSMs and/or DMMs.

After the elements of the first two domains are specified, the threat scenario can be assembled. The composition of a valid scenario without any circular logic in the building process can be assured by choosing the elements according to the sequence indicated by the MDM. Group 2 in figure 9.9.3 contains the relevant domains for this. Each selection affects the elements in the following domains; they are reduced to the ones consistent with the chosen scenario. When at least one element of each domain is settled (multiselection of some elements is possible, such as in the *tool/weapon* domain), a structurally consistent scenario is completed. In addition to the scenario, it is important to know the attacker's way through the airport. Based on this information, scenario-specific security measures can be deduced. Possibilities are greatly reduced by specifying the use case (group 1). Additional choices have to be made in group 3 (the dependencies between threat scenarios and the airport layout).

The remaining areas of the MDM contain security measures addressing single elements of the scenario (group 4) and information about the specific airport's security infrastructure (group 5). The information from these two parts of the MDM is needed to evaluate the airport's capacities to address the threat.

Results

An important result was a well-documented structure of the system and the interrelations of its elements—already achieved during the data acquisition phase. This clarified the definitions shared by all the participants.

Systematic data acquisition provided the basis for a structured assessment of threat scenarios. The system of airport security was too large for reasonably tracking the connection of each desired pair of scenario elements in the matrix, given the required level of detail. For this reason, we developed a tool for facilitating the data access. A scenario builder draws on the data gathered in the MDM and guides the user through the process of building a plausible scenario. It provides the sequence in which the elements need to be specified: Elements can only be chosen if they are consistent with the prespecified aspects of the scenario. Thus, it is impossible to assemble structurally inconsistent scenarios when working with the builder. Furthermore, after completing a scenario, the builder automatically indicates which security activities and technologies address elements of the respective scenario. The tool offers intuitive interaction with the complex structure, making the broad space of all structurally consistent scenarios accessible.

In planning airport checkpoints while taking possible future threats into account, it is desirable to account for as many scenarios as possible. Because the manual creation of scenarios is time consuming, the scenario builder has been automated, permuting through all possible element combinations and consequently producing all of the structural possibilities in the scenario design space.

Analyzing these data gave us hints concerning weak spots in the existing structure: Scenario clusters with few security technologies and activities addressing them might not be well protected. However, scenarios addressed by a large number of security measures might hint at possible redundancies in the airport layout. Such an analysis serves as a basis when testing the implementation of alternative techniques and layouts: If a poorly protected scenario cluster is addressed by new processes or technologies, then new measures seem appropriate.

References

Cole, Mara, and Andreas Kuhlmann. 2010, June. *Preparing Today's Airport Security for Future Threats—A Comprehensive Scenario-Based Approach.* Proceedings of the 12th annual conference of the Finland Futures Research Centre, Turku, Finland.

Cole, Mara, Andreas Kuhlmann, and Oliver Schwetje. 2009, June. *Aviation Security—A Structural Complexity Management Approach*. Proceedings of the 13th Air Transport Research Society World Conference, Abu Dhabi, United Arab Emirates.

Maurer, Maik, Wieland Biedermann, Andreas Kuhlmann, and Thomas Braun. 2009, October. *The 2-Tupel-Constraint and How to Overcome It*. Proceedings of the 11th International Design Structure Matrix Conference, Greenville, SC.

Maurer, Maik, Wieland Biedermann, Mara Cole, John D'Avanzo, and Dirk Dickmanns. 2009, December. *Airport Security: From Single Threat Aspects to Valid Scenarios and Risk Assessment*. Proceedings of the 1st annual Global Conference on Systems and Enterprises, Washington, DC.

Example 9.10   4G Mobile Phone LSI Chip Design

Contributors

Tsuyoshi Koga, Akihiro Hirao, and Kazuhiro Aoyama
Department of Systems Innovation, University of Tokyo

Yoshiharu Iwata
Center for Advanced Science and Innovation, Osaka University

Problem Statement

The design of large-scale integration (LSI) chips in the semiconductor industry has entered a phase of major change. As the spacing between transistors narrows, it is predicted to reach a physical limit. A major focus of R&D efforts today is how to find better structures for LSI chips, such as System-on-a-Chip (SoC) or System-in-a-Package (SiP) architectures. Hence, a new method of supporting design decisions in the initial design stage is strongly desired. To help improve LSI chip design, we built an MDM model to increase understanding of the initial design and engineering processes.

Data Collection

Through discussions with industrial design engineers in Committee No. 177 on System Design and Integration in the Japan Society for the Promotion of Science, we captured 170 typical design parameters (e.g., memory capacity) and 100 design tasks, which are related to parameters through equations (e.g., memory capacity is calculated from the total area and spacing between transistors). The dependencies between parameters therefore come from the equations. We represented the design system of a LSI chip based on these equations, which consist mainly of four domains: computing, thermal, electrical, and spatial.

Model

Figure 9.10.1 illustrates the structure of the overall MDM model, which contains four DSMs. The system-based DSM represents the dependencies between subsystems. The component-based DSM represents the dependencies between product components. The parameter-based DSM represents the dependencies between parameters. The task-based DSM represents the dependencies between design tasks. The relationships among subsystems, components, parameters, and design tasks are defined in the DMMs.

Figure 9.10.2 shows the task-based DSM for LSI chip design. Design tasks in this case are the mathematical or empirical relationships (such as equations) between parameters.

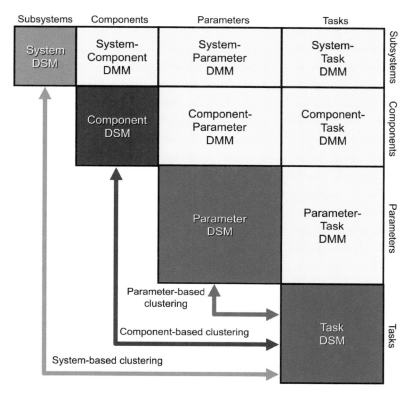

Figure 9.10.1
Structure of the overall MDM model.

The off-diagonal cells in the matrix identify the number of interactions between design tasks via the parameters. The parameter design process can be planned based on this matrix.

This matrix is somewhat difficult to interpret because the dependencies are widely distributed and often bidirectional. It lacks coherence because this initial DSM of mathematical relationships does not specify any parameter design clusters or sequence. However, we can analyze this DSM to reveal a structure, which indicates how to efficiently execute the design process. We begin the analysis by identifying design clusters, and then the design process is obtained.

Results

Figure 9.10.3 shows the result of clustering the process DSM based on an understanding of the overall LSI structure and four types of domain knowledge in the MDM. Each cluster is a group of tasks based on the architecture of subsystems, components, and

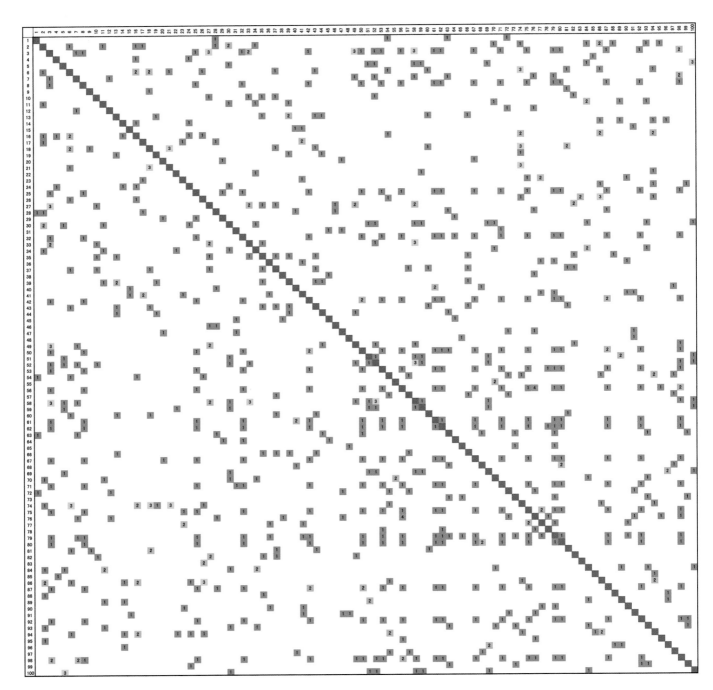

Figure 9.10.2
Process (task) DSM of 100 design tasks and their interactions.

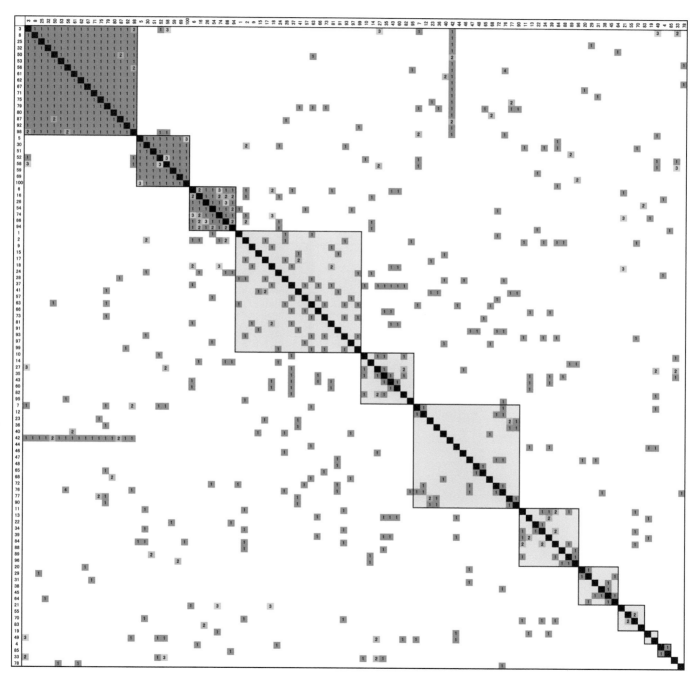

Figure 9.10.3
Process (task) DSM clustered on the basis of other domains in the LSI chip architecture MDM.

parameters, with dependencies between design parameters collected inside the clusters. Based on the DMMs, three different kinds of clustering algorithms were used for the task-based DSM: (1) subsystem-based (hierarchical) clustering, (2) component-based (structural) clustering, and (3) parameter-based (functional) clustering. Three different sets of clusters (hierarchical, structural, and functional) were thus obtained. The LSI designer reviewed these results and selected good clusters. Figure 9.10.3 shows one of these clustering results.

Figure 9.10.4 shows a clustering result with 170 parameters, resulting in a design sequence. To obtain this result, the clustered, parameter DSM was sequenced two times, first by sequencing the clusters and second by sequencing the parameters within each cluster. Figure 9.10.4 therefore suggests an overall design process for the LSI chip, with three main findings: (1) the logic chip should be designed before the dynamic random

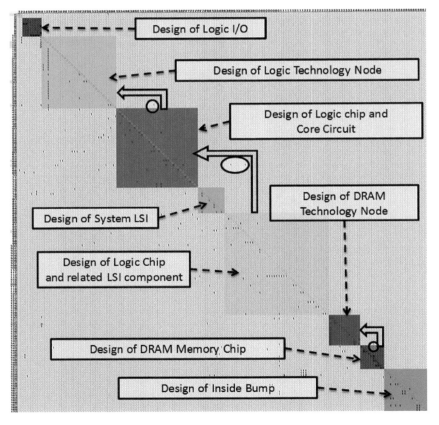

Figure 9.10.4
Clustered and sequenced parameter-based DSM, suggesting the overall LSI design process at the parameter level.

access memory (DRAM), (2) the technology node (which sets the transistor spacing) should be determined before the logic chip and DRAM design, and (3) the inside bump (an interface between logic and memory chip) should be designed last.

References

Hirao, Akihiro. 2009, March. *Design Process Planning for Development of Complex Product System*. Bachelor's thesis, University of Tokyo, Tokyo, Japan.

Hirao, Akihiro, Tsuyoshi Koga, and Kazuhiro Aoyama. 2010 July. *Planning Support of Initial Design Process Based on Clustering and Ordering of Tasks—Design Example of an Integrated Circuit*. Proceedings of the 12th International Design Structure Matrix Conference, Cambridge, England, pp. 83–96.

Koga, Tsuyoshi, Wataru Ono, Akihiro Hirao, and Kazuhiro Aoyama. 2010, October. *Structuring of Early Process of Product Development Considering Order and Dependencies between Design Tasks*. Proceedings of the JSME D&S conference, 2310(CD-ROM), Japan Society of Mechanical Engineers, Tokyo, Japan.

Example 9.11   Audi AG Body-in-White Development

Contributor

Matthias Kreimeyer
Technische Universität München

Problem Statement

In the context of a research project on the improvement of communication among the different design and simulation departments concerned with the design of the body-in-white at Audi AG, Germany, the company's overall design process was analyzed to reveal the tasks and business objects (work products) that guide the development process at the organizational interfaces. Overall, a balanced improvement incorporating process, organizational, and information technology aspects was desired, so an MDM approach was chosen.

Data Collection

The MDM was exported from a process model built in event-driven process chain notation using the ARIS Toolset 6.1 by IDS Scheer AG. Each task represents a work package of four to six weeks of effort for one organizational unit. The process model was built based on 68 interviews with various staff involved in the process, involving the domains shown in the meta-MDM in figure 9.11.1. For each interview, individual process models were built and later consolidated in a series of workshops to form the overall process model. In addition, the organizational structure was added to complete the model as well as the checklists for necessary deliverables that were used for each milestone. The data on milestones were rather incomplete and were finally omitted. For the modeling of tasks and business objects (e.g., the "crash simulation results"), a denomination scheme was used to designate the responsibility of the involved personnel (e.g., "support," "do," "coordinate," etc.) or the type of task ("make concept," "develop," etc.).

Model

OR gates were modeled explicitly, representing points in the process where the flow of information is either split or joined. Only 54 explicit decisions were modeled (i.e., decisions that were taken actively during the design process). To generate a simple model, AND decisions were not explicitly modeled, and XOR decisions were represented only as OR.

	Tasks	Business Objects	Org. Units	Milestones	IT Systems	OR Gates
Tasks		T generates BO ID 1				T generates BO ID 2
Business Objects	BO is input for T ID 3			BO is necessary to reach M ID 4		BO is input for T ID 5
Organizational Units	OU is responsible for T ID 6					
Milestones						
IT Systems	IT supports T ID 7				IT has interface to IT ID 8	
OR Gates	BO is input for T ID 9	T generates BO ID 10				OR precedes OR ID 11

Figure 9.11.1
Meta-MDM used for process analysis (IC/FBD convention).

Such decision points occur between tasks and business objects (e.g., when a task makes a decision that results in different business objects and vice versa). This leads to the four DMMs (IDs 2, 5, 9, and 10) shown in figure 9.11.1. Additionally, one decision can lead to another, represented as a DSM as shown in the same figure (ID 11). To represent these OR gates, therefore, each decision point was modeled as an individual entity of a new domain ("OR Gates") and related with the appropriate relationship types, either "T generates BO" or "BO is input for T," according to where the decision point was inserted.

More generally, decision points, modeled as OR gates in this case, can be represented as a separate domain that does not have an impact on the relationship type. The example in figure 9.11.2 shows how two business objects BO 1 and BO 2 lead to two tasks T 1 and T 2 (or any combination thereof, as shown by the joining OR gate OR 1 and the splitting OR gate OR 2). These dependencies are shown in the MDM on the right hand side, which represents the three domains and the appropriate relationship types as shown in figure 9.11.1.

The MDM shown in figure 9.11.3 was exported using a standardized export function from the process modeling tool that delivers various lists. These lists were, in a second step, used to build the DMMs and DSMs. Overall, 160 tasks, 134 business objects, 14

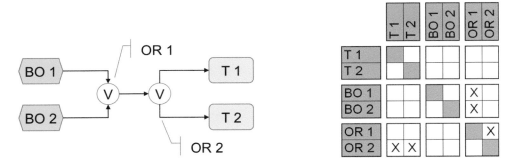

Figure 9.11.2
Task DSM derived from business objects and OR gates (IC/FBD convention).

organizational units, and 27 IT systems comprise the MDM, which was validated through several workshops and analyses.

The native data were used to derive the task DSM shown in figure 9.11.4 by following the flow of information via the business objects exchanged by the tasks and via the decision points (two at most between any two tasks) within the process. If, for example, the task "coordinate aeroacoustics" is followed by a business object "results for aeroacoustics simulation," which leads to the task "support development of structure," then the first task was linked directly to the second task in the derived DSM. Intermediate OR gates were treated similarly (Kreimeyer and Lindemann 2011). The resulting DSM was randomly cross-checked with engineers involved in the process to ensure that the aggregation did not bring about false results. In theory, other DSMs could be calculated (e.g., via common IT systems), but these were not regarded here because the derivation procedure generally yielded dense DSMs that did not provide much insight. (This is a common problem when deriving a DSM for a domain with fewer nodes.)

Results

To obtain the core drivers of the process, five different complexity metrics were applied to the derived task DSM. First, the in-degree and out-degree regard the immediate context of each task and count its incoming and outgoing interfaces, respectively. Thus, the degree (the sum of the in- and out-degrees) identifies tasks that demand high coordination effort and that are critical sources or sinks of information. The degree distribution of all tasks can be plotted as a frequency histogram, as in figure 9.11.5. This shows that the process depends largely on tasks that are only minimally connected, while there are a few key tasks with a degree of around 10 to 12, and there are two nodes with degree 30 and 32, respectively. These two tasks act as major information sinks and thus as hubs in the immediate context of the process.

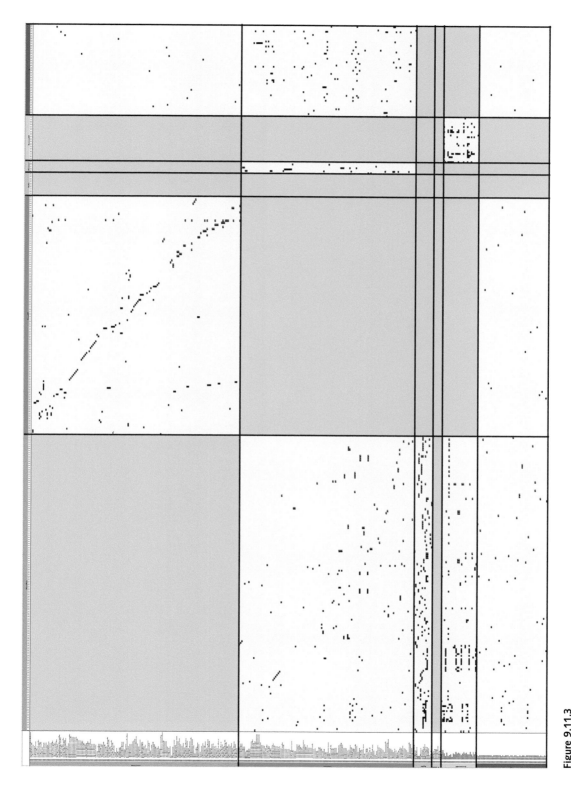

Figure 9.11.3
Overall MDM with native data (empty matrices shaded grey).

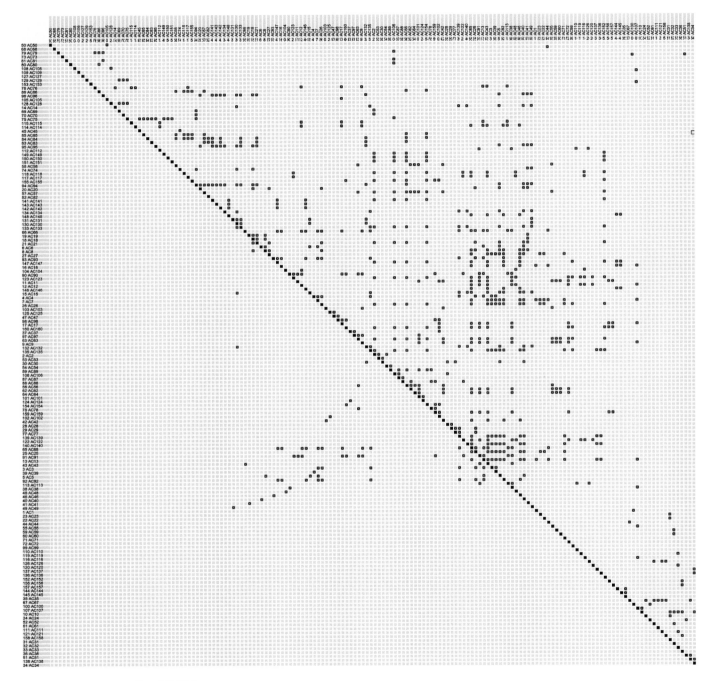

Figure 9.11.4
Task DSM derived from business objects and OR gates (IC/FBD convention).

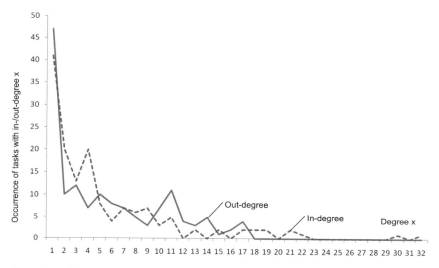

Figure 9.11.5
Degree distribution for in-degree and out-degree.

The second complexity metric, reachability (outgoing and incoming), specifies the path length by which a task can reach every other task in the process. It thereby extends the degree measure to the whole process and not just the immediate vicinity of a task.

Third, the snowball and forerun factors regard the outgoing and incoming hierarchies of each task, respectively (i.e., the nodes that can be attained from each task directly or via n intermediate tasks). However, they do not simply assess how many other tasks can be reached; they also assess the distance to each task in the hierarchy, thus taking into account that tasks which are farther away have less impact. Hence, they normalize the reachability measures.

Fourth, relative centrality counts the number of geodesics (shortest paths) between any pair of tasks that pass via a given task, therefore assessing how many information channels go via any given task. This metric helps to assess how much a task contributes to architecting a system, as a more central task will have a greater influence on how information is processed.

Fifth, the occurrence of iterations is assessed by regarding how many cycles in the DSM include each task. As such, the concept is similar to that of the relative centrality, but focusing on how much each task contributes to rework cycles in the process, thereby being more relevant to the problem-solving process. Although these numbers may seem large, they represent the occurrence of each task in all cycles from length two (one task to the next and back) to the longest cycle. These results are in line with the high centrality of the three tasks shown: The more central a task is to the process, the more rework cycles

		1		2		3		4	5
		Out degree	In degree	Outgoing reachability	Incoming reachability	Snowball-factor	Forerun-factor	Relative centrality	Number of cycles per node
AC 43 - Setup simulation model for crash		3	32	105	108	25.8	62.7	396	145,754
AC 65 - Coordinate simulation of crash		16	30	109	111	51.9	62.4	3233	205,467
AC 91 - Coordinate simulation of passenger safety		12	15	108	112	47.3	50.7	1190	156,927

Top outlier	Second outlier	Third outlier

Figure 9.11.6
Complexity metrics for three tasks.

will typically go across it. Figure 9.11.6 displays results for three selected tasks. The shading indicates that a task was among the top three values (i.e., a "structural outlier").

Coordinate simulation of crash (task 65) is by far the most important task of the process, being among the top three for almost all metrics. This is in line with Audi's strategy for vehicles, as crashworthiness is one of the most important properties a car is optimized toward. The task is well embedded in its immediate vicinity, as both a high in- and out-degree show. The high in-degree refers, in fact, to the need to collect a lot of different business objects to build a crash simulation model. Also, the crash model is later used for other simulations and therefore forwarded to other tasks, as the high out-degree shows. At the same time, the results from this task are used to improve the design for the car body, hence its high centrality. However, the factual setup of the simulation model (i.e., not the coordination of this process) makes use of even more inputs, as the in-degree shows, while having relatively few outputs. This task collects information from all across the process, which results in the highest possible value for the forerun factor.

As a result, a process improvement project was focused to work, in a first step, on raising the efficiency of simulation models, as these business objects showed to be the most central in the process. Using the MDM, the stakeholders involved in building the various business objects (e.g., simulation models) could be derived as a starting point for this improvement project.

The separation of the different domains in the process analysis provided two advantages versus using only a DSM. First, the MDM could simply be generated by exporting various DSMs and DMMs out of a standard process modeling tool (ARIS Toolset 6.1), and thus the native dependency information could be generated in the way that engineers at Audi AG were used to, thereby making sure the information was well understood, correct, and consistent. Second, the analysis of the Task DSM could always be traced back to involved business objects, hence facilitating the interpretation of the results (e.g., the task "Coordinate simulation for crash" is involved in many rework cycles that, to a large extent, depend on a single business object, although it is linked to several).

References

Kreimeyer, Matthias, Stefanie Braun, Matthias Gürtler, and Udo Lindemann. August, 2009. *Extending Multiple Domain Matrices to allow for the Modeling of Boolean Operators in Process Models.* International Conference on Engineering Design—ICED'09, The Design Society, Stanford, CA.

Kreimeyer, Matthias, and Udo Lindemann. 2011. *Complexity Metrics in Engineering Design.* Berlin: Springer.

Example 9.12   U.S. Air Force MAV Development

Contributors

Jason Bartolomei, Richard de Neufville, Daniel Hastings, and Jennifer Wilds
Massachusetts Institute of Technology

Problem Statement

The U.S. Air Force Research Laboratory (AFRL) is the U.S. Air Force's leading organization dedicated to the discovery, development, and integration of war fighting technologies for air, space, and cyberspace forces. The AFRL was tasked to develop a miniaturized unmanned air vehicle (MAV) for U.S. Special Forces (figure 9.12.1). This quick reaction project integrated several technologies under development within the AFRL, as well as technologies developed by industry and academia. To understand this complexity, this project applied MDM analysis using product, organization, and process DSMs and their corresponding DMMs. The model used a time-series data set to capture the dynamic complexity of a product development system. This allowed examination of the impact of engineer turnover within the design organization, the effects of changing requirements on the design, and the design evolution. Analysis identified platform and modularity opportunities in the design by allowing system engineers to explore the sources and effects of design changes in the product development system.

Data Collection

Over a two-year period, Jason Bartolomei (a U.S. Air Force officer and PhD student in MIT's Engineering Systems Division) observed the MAV project underway at the AFRL. While constructing the MDM model for the MAV project, Bartolomei found that much of the information surrounding the system resided not in technical documentation but rather in the stories of the people in and around the system. Transparency of assumptions and traceability of sources were vitally important due to the qualitative nature of the system's data, the scope of the system that transcended one person's direct knowledge and expertise, and the system complexity measured by the number of components and interactions.

As such, Bartolomei developed and used a data collection and multidomain modeling technique suitable for the types of qualitative knowledge data surrounding the development of complex engineering systems. The method follows an iterative, systematic process that translates system information collected through interviews, observation, and documentation into an Engineering Systems MDM (ES-MDM). The modeling process is comprised of the following steps:

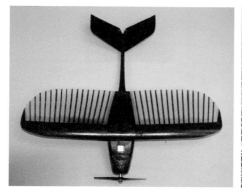

Figure 9.12.1
A miniaturized unmanned air vehicle (left) and its ground control station (right).

1. **Identify the system of interest**   The ES-MDM framework consisted of six DSMs and 15 DMMs as shown in figure 9.12.2. (The interdomain relationships are undirected, so the ES-MDM is symmetric, and the 15 DMMs above the diagonal are the same as the 15 below.) The six DSMs addressed the following domains:

   - **System Drivers (Environmental) Domain**   A variety of factors including regulatory agencies, other military organizations, various technologies, military acquisition system, congressional budgets, and others. Example components in this DSM include changing threats ("red force"), rapidly advancing technologies, as well as changing interfaces to friendly ("blue force") technologies and tactics that the MAV needed to interact with to perform user-defined missions.

   - **Stakeholders (Organization) Domain**   AFRL managers, engineering, and technical support staff, as well as several subcontractors responsible for the development and testing of technical subsystems, plus a variety of external stakeholders.

   - **Objectives Domain**   Purposes and goals of the system; here, the design and development of MAV prototypes that meets customer needs on schedule and within costs.

   - **Functions Domain**   A decomposition of the objectives into a hierarchy of functions and subfunctions.

   - **Objects (Product) Domain**   The MAV system, subsystems, and components, including a laptop computer, antennae, backpack, as well as fabrication equipment used in production.

   - **Activities (Process) Domain**   Processes, activities, and tasks involved in the design, development, and management of the MAV system.

	System Drivers	Stakeholders	Objectives	Functions	Objects	Activities
System Drivers	The list and interactions of exogenous factors that act or acted on by the system	Relates the stakeholders that act on exogenous variables	Relates the objectives that act on exogenous variables	Relates the functions that act on exogenous variables	Relates the technical components that act on exogenous variables	Relates the activities that act on exogenous variables
Stakeholders	Relates the exogenous variables that act on system stakeholders	The list and interactions of the human entities within the system	Relates the objectives that act on stakeholders	Relates the functions that act on stakeholders	Relates the technical components that act on stakeholders	Relates the activities that act on stakeholders
Objectives	Relates the exogenous variables that act on system objectives	Relates the stakeholders that define or contribute to the system objectives	The list and interactions of combined purposes and goals of the system	Relates the functions that act on or relate to system objectives	Relates the technical components that act on system objectives	Relates the activities that act on system objectives
Functions	Relates the exogenous variables that act on system functions	Relates the stakeholders that act on system functions	Relates the objectives that are decomposed into system functions	The list and interactions of functions of the system	Relates the technical components that are traceable to system functions	Relates the activities that act on system functions
Objects	Relates the exogenous variables that act on system technical components	Relates the stakeholders that act on the technical components of the system	Relates the objectives that act on or constrain technical components	Relates the functions that are allocated to technical components	The list and interactions of technical components of the system	Relates the activities that act on technical components
Activities	Relates the exogenous variables that act on the system activities	Relates the stakeholders that engage in or act on the activities of the system	Relates the objectives that act on or constrain system activities	Relates the functions that are allocated to system activities	Relates the technical components that act on system activities	The list and interactions of activities of the system

Figure 9.12.2
The Engineering Systems MDM (IR/FAD convention).

2. **Define objectives for analysis**   See the "Problem Statement" section. Because time-series information for each of the domains was important, multiple ES-MDMs would be needed to represent different time periods.

3. **Collect data**   Qualitative, social science methods for eliciting data through interviews are central to constructing the ES-MDM. Subject matter experts were interviewed with open-ended questions, and recorded interviews were transcribed. In addition, pertinent documentation describing the system was collected, including technical data used for computational models, engineering drawings, e-mails, and program presentations. Data were collected over 24 months (December 2004–December 2006) and represent all 46

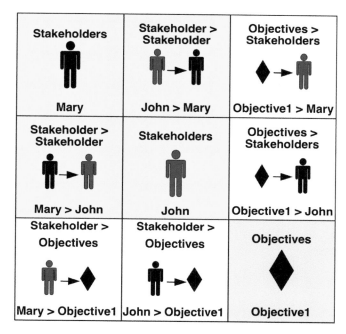

Figure 9.12.3
Placing the coded data into the ES-MDM (step 5).

months of the MAV project duration (February 2003–December 2006). All together, several thousand pages of interview transcripts, program documentation, and other data were collected.

4. **Code the data**   Tag the data as they pertain to the nodes, relations, and attributes of the six domains of the ES-MDM.

5. **Organize the coded data in the ES-MDM**   Figure 9.12.3 demonstrates this step for the interview transcripts.

6. **Examine the model for missing/conflicting data**   Each element of the model can be referenced to raw data (interviews, documentation, etc.). Experts were invited to review and verify the data and the model.

7. **Resolve missing data**   Take action to resolve conflicts. This was done through additional interviews, reviews of the raw data, and other similar actions.

8. **Perform analysis**   Various quantitative analytical methods are available to examine the system structure and behavior.

9. **Iterate**   Modelers are likely to perform several iterations of the methodology in the analysis of a complex system.

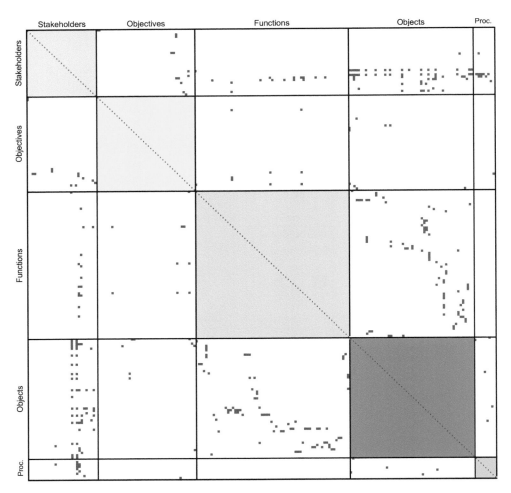

Figure 9.12.4
The full ES-MDM (at time 1).

Figure 9.12.4 shows the overall MAV ES-MDM, and figure 9.12.5 shows the stakeholders DSM portion of the ES-MDM in greater detail.

Results

One important question facing project leadership was, "Should we be managing this project at the subsystem or component level?" Insights into this question drive organizational structure, resource decisions, and process development. To explore this question, we computed a common network metric, betweenness centrality (BC), which measures

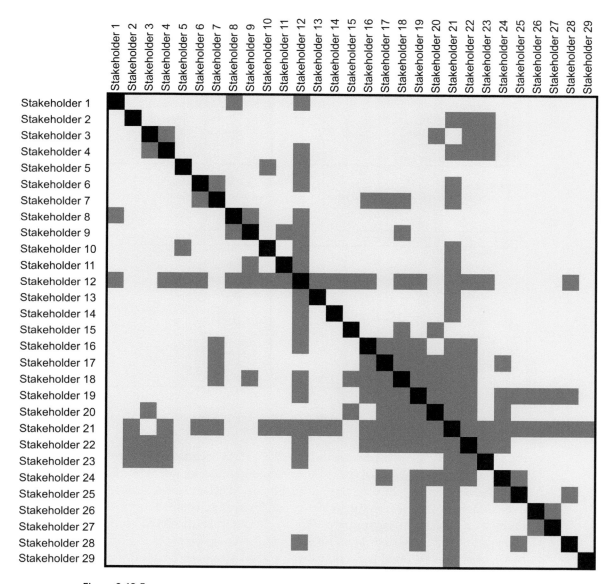

Figure 9.12.5
Stakeholder DSM portion of the ES-MDM.

Rank	Objects DSM Only	BC
1	Engine Subsystem	373
2	Ground Station Transmitter	272
3	Control Subsystem	244
4	Ground Station Subsystem	212
5	Ground Station Software	197
6	Actuator #1	154
7	Wing Subsystem	138
8	Battery Connectors	135
9	Ribs	127
10	Wing Composite Structure	103

Rank	Entire ES-MDM	BC
1	Autopilot Subsystem	1977
2	Communication Subsystem	1822
3	Ground Station Subsystem	1749
4	Air Vehicle	1388
5	Wing Subsystem	1299
6	Battery Subsystem	1013
7	Fuselage Subsystem	1008
8	Ground Station Software	992
9	Control Subsystem	967
10	Fuselage Structure	967

Figure 9.12.6
Betweenness centrality (BC) for Product component (objects) only (left) and all ES-MDM elements (right).

Rank	MAV-PD Social Network	BC
1	PMWJ	500
2	STCC	199
3	PMBI (MAV-PD PM 3)	84
4	SPOMD	55
5	SPOKE	45
6	SPOGR	44
7	KTRDM	40
8	STYA	22
9	STSP	20
10	PMFC	15

Rank	MAV-PD Entire Network	BC
1	PMWJ	10973
2	KTRDM	3680
3	KTRNM	1972
4	STCC	1557
5	PMBI (MAV-PD PM 3)	1373
6	KTRRC	1004
7	KTRTT	618
8	KTRBR	390
9	SPOMD	293
10	STYA	275

Figure 9.12.7
BC for stakeholders only (left) and stakeholders rankings with respect to all ES-MDM elements (right).

the number of times an element within the ES-MDM occurs on the short path connecting two other elements. We used it to compare the top ten components in the Objects Matrix with the top ten elements in the entire ES-MDM. The results, shown in figure 9.12.6, indicated that subsystems have greater connectivity than components in the larger system. From a management perspective, this suggested that the MAV program should be managed at the subsystem level rather than the component level. For systems like the MAV, organizing around subsystems was a good approach because the product system integrated several commercially available subsystems. By managing at the subsystem level of complexity, the program could optimize the allocation of manpower and better facilitate systems integration.

The same analysis was used on the Stakeholders DSM. In social network analysis, BC is associated with the control of information: Stakeholders with higher BC have greater influence on a social network. Figure 9.12.7 shows the results, at time 3, where each acronym "XXXX" represents a particular individual. In the social domain, the highest

		Time 1	Time 2	Time 3	Time 4	Time 5
PMWJ	Degree Centrality	38	46	61	53	8
	BC	3643	5427	11836	10331	1769
STCC	Degree Centrality	16	21	28	23	4
	BC	820	866	1667	3501	241
MAV-PD Avgs	Degree Centrality	4.9	4.9	4.7	5.0	3.8
	BC	238	258	280	296	339

Figure 9.12.8
Centrality measures for MAV stakeholders over time.

ranked stakeholders were PMWJ and STCC, the chief engineer and the customer, respectively. PMBI was the program manager; SPOMD, SPOKE, and SPOGR were contract managers; and KTRDM was the lead contractor responsible for the ground station and production. These results were as expected. The rankings changed, however, when the analysis was expanded to include the functional, technical, and process domains, as shown on the right side of figure 9.12.7. For example, subcontractor KTRDM's ranking surpassed that of the program manager PMBI. The ES-MDM thus revealed that KTRDM, a seemingly less important stakeholder when looking at the social network, was far more connected and had greater influence in the overall MAV engineering system. The ES-MDM provided a means to better understand the roles of stakeholders involved in the system.

The ES-MDM was also used to examine how stakeholder characteristics changed over time. We compared stakeholders PMWJ and STCC's degree centrality (the count of connections for a node) and BC over several six-month intervals (figure 9.12.8). The analysis compared these centrality measures with those for their replacements at time 5. The metrics for both PMWJ and STCC grew over time and were always much larger compared with those of the average individual within the MAV system. At time 5, however, both PMWJ and STCC were removed and replaced with new agents. The replacements had significantly smaller centrality measures, an indication that they were not as well connected in the system. In this case, the ES-MDM and analysis can be used as a tool to identify the downside risks associated with personnel changes. For the MAV program, the changeover in personnel correlated with a decline in project performance. A benefit of the ES-MDM is that it gives new personnel a means to engage more deliberately in the system by revealing important connections within and across domains.

We also analyzed the entire MAV development system over time. As shown in figure 9.12.9, the size of the MAV network and the density of the relationships changed over time. These changes are not surprising because frequent organizational and technology changes were well documented. It is interesting to note the difference in network metrics at time 5. The degradation in the number of relations far exceeds that in the number of nodes. The average degree <k> and clustering coefficient is lower compared with those

MAV ES-MDM	# of nodes, n	# of links, m	Avg. degree, $<k>$	Avg. path length, l	$\dfrac{\log\ n}{\log\ <k>}$	Clustering coeff., $C^{(1)}$	Clustering coeff., $C^{(2)}$	$<k>/n$
Time 1	184	884	4.87	3.61	3.29	0.29	0.21	0.03
Time 2	196	956	4.93	3.67	3.31	0.31	0.20	0.03
Time 3	262	1232	4.70	3.62	3.60	0.32	0.20	0.02
Time 4	223	1106	4.96	3.67	3.38	0.30	0.19	0.02
Time 5	206	778	3.77	4.44	4.01	0.22	0.13	0.02
Stakeholders DSM Time 3	47	328	6.97	2.05	1.98	0.68	0.33	0.15
Objects DSM Time 3	52	226	4.34	2.93	2.69	0.32	0.22	0.08

Figure 9.12.9
Various complexity metrics computed over five time periods.

in other periods, and path length seems much longer. This observation prompted a reexamination of the system for insights as to what might have happened.

At time 5, the changes in the network metrics reflect the loss of PMWJ and STCC. Their replacements from outside organizations had no experience with the MAV. This disrupted the cohesion of the MAV team, as management changed from the flat structure created by PMWJ and STCC's social connections into what seemed to be a classic military stovepipe. The time 5 ES-MDM shows a longer average path length and a smaller average clustering coefficient, which supports the qualitative data indicating significant structural changes when PMWJ and STCC left the project. Efforts to develop new MAV prototypes then rapidly diminished. Within six months, the project lost its capacity to develop MAV prototypes. The project objectives then changed to develop a small subset of the original system.

The ES-MDM holds promise as an industrial tool for systems engineers. The information used to build common system-level models and products such as QFD and DoDAF can be captured in a fully attributed ES-MDM. Plus, the ES-MDM captures information about the social and environmental domains important to systems engineering projects. Last, a collection of ES-MDMs can represent time-series information about a project.

References

Bartolomei, Jason E. 2007, June. *Qualitative Knowledge Construction for Engineering Systems: Extending the Design Structure Matrix Methodology in Scope and Procedure*. PhD thesis, Engineering Systems Division, Massachusetts Institute of Technology, Cambridge, MA.

Bartolomei, Jason E., Daniel E. Hastings, Richard de Neufville, and Donna Rhodes. 2012. Engineering Systems Multiple Domain Matrix: An Organizing Framework for Modeling Large-Scale Complex Systems. *Systems Engineering* 15 (1):41–61.

Bartolomei, Jason E., Susan S. Silbey, Daniel E. Hastings, Richard de Neufville, and Donna Rhodes. 2009, June. *Qualitative Knowledge Construction for Engineering Systems: Bridging the Unspannable Chasm between Social Sciences and Hard Sciences*. Proceedings of the International Conference on Engineering Systems, Massachusetts Institute of Technology, Cambridge, MA.

Example 9.13   Kalmar Industries Supplier Network

Contributors

Mike Danilovic
Halmstad University

Mats Winroth
Chalmers University of Technology

Problem Statement

Kalmar Industries produces heavy-duty materials handling equipment such as reach-stackers that are used in port and transportation operations. To deliver anticipated large customer orders of reach-stackers in a limited time frame, Kalmar worked to strengthen and intensify its collaboration with three major suppliers, Hiflex, Euromaster, and Kone, in a joint, co-located industrial network. The major challenge was to design the collaborative and information exchange processes between the four companies.

Data Collection

Based on the extended Miltenburg framework (1995), we formulated a large MDM model of the industrial network surrounding the reach-stacker development and production. Three major aspects were modeled for each of the four companies involved—the competitive analysis focusing on market and customer requirements, the decision criteria regarding design of the manufacturing and management system, and layout of the manufacturing process—resulting in the 216 x 216 matrix shown in figure 9.13.1. Data for the MDM model came from interviews with managers at the four companies. The number and shading in each off-diagonal cell represents the strength of the dependency: 3 (red) = very strong, 2 (pink) = medium, and 1 (yellow) = low. All three domains are based on a static analysis, but there is a dynamic relationship between those three. They influence each other within and between each company.

Each of the three aspects of the process involves numerous interactions among Kalmar, the system integrator (SI), and the suppliers. The SI has to analyze the market situation and customer requirements. In this process, the present manufacturing systems influence the analysis of decision areas, which is fed back to the design of the production system. There is also a process of synchronization between the SI and the preferred suppliers regarding market situation and customer demands, as well as a negotiation of how the suppliers should organize their production systems according to what the SI is capable of doing on its own and what part of the supplier organization should be relocated within proximity of the SI. Third is a process of adaptation within and between each supplier.

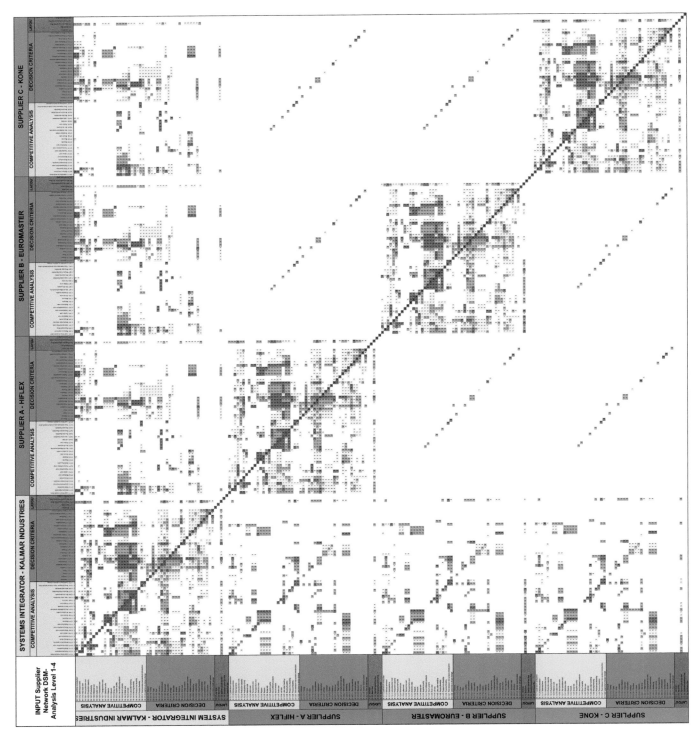

Figure 9.13.1
Original MDM of the entire corporate network.

They have to decide how to respond to the demands, what part to relocate, and how to develop new organizational routines to handle the daily activities, not only in their own corporation but also within the entire network.

Model

The initial MDM model shown in figure 9.13.1 represents grouping according to the four corporate entities.

Results

We initially performed a clustering analysis at the corporate network level, analyzing the entire MDM as if it were a large DSM. We used the clustering algorithm of McCormick et al. (1972), followed by some further, manual adjustments. Figure 9.13.2 shows the result, where we identified five clusters along the diagonal. We allowed these clusters to overlap, designating the common element in two clusters as a linking element. Figure 9.13.3 zooms in on cluster 3. Here we can see detailed interdependencies among all four corporations (SI and suppliers A, B, and C) that span all three aspects of the Miltenburg framework. Although it had been clear that the SI needed to work closely with each supplier, this analysis revealed the importance of the suppliers also working closely with each other.

In the next step, we treated the large MDM as a combination of DSMs and DMMs, as shown in figure 9.13.4. DMMs 1 and 4 show analysis between Kalmar Industries and Hiflex, DMMs 2 and 5 show analysis between Euromaster and Kalmar Industries, and so on. DMMs 7–12 show analyses between suppliers. The final analyses we conducted were on the separate DSMs and DMMs, focusing on dependencies among the four companies.

For example, figure 9.13.5 shows DMM 4, which has been clustered to show three major areas in which detailed information flows create interdependencies between Kalmar Industries and Hiflex. (The same procedure was used for each of the other DSMs and DMMs.) These three clusters were identified by visualization and manual evaluation of the elements. In order for the network to function efficiently, it is crucial that each company understand what kind of information different stakeholders need and how each other's competitive advantage is influenced by strategic decisions, actions, and interconnectivities. The combination of DSM and DMM approaches used in this research explores and enables synchronization, within and between companies, regarding aspects of their intra- and interfirm competitive situation, decision criteria, and production system layout. The participative approach that we have chosen, involving people from all companies of the network, involves several functions and strategizing processes in the analysis, reveals assumptions regarding the market and competitive situation, and explores the need

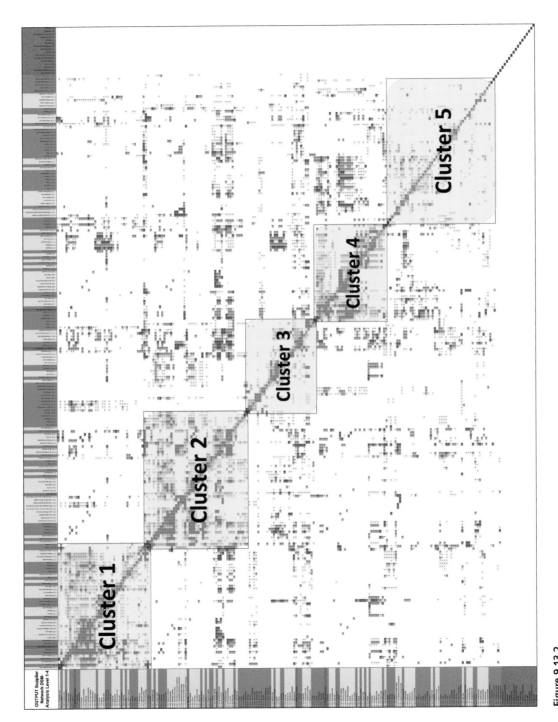

Figure 9.13.2
Initial analysis of the overall corporate network via clustering.

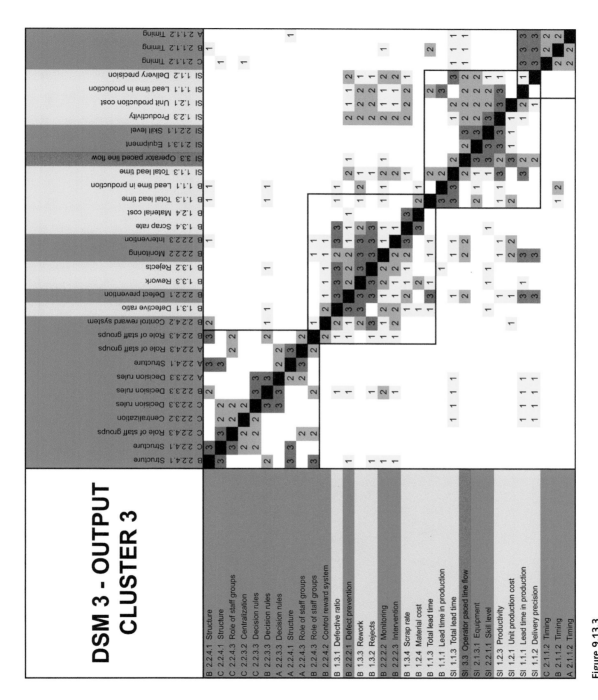

Figure 9.13.3
Detailed view of cluster 3 from figure 9.13.2.

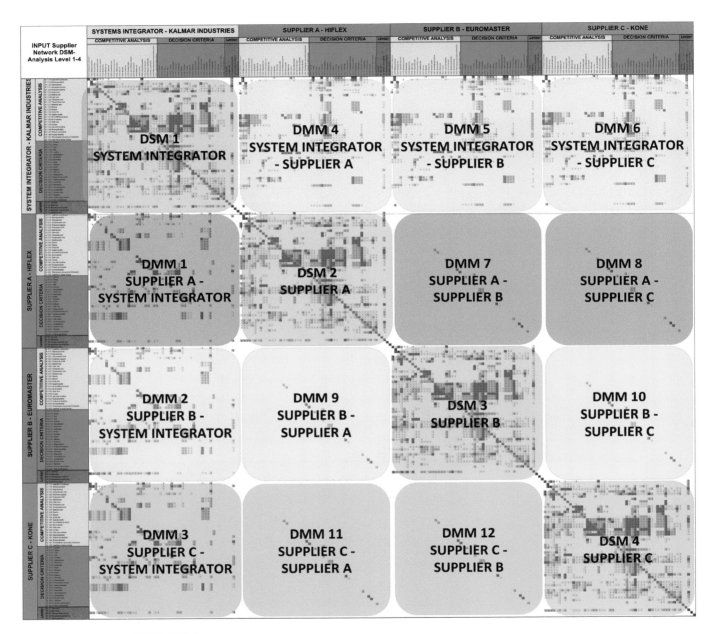

Figure 9.13.4
Parsing the MDM into DSMs and DMMs.

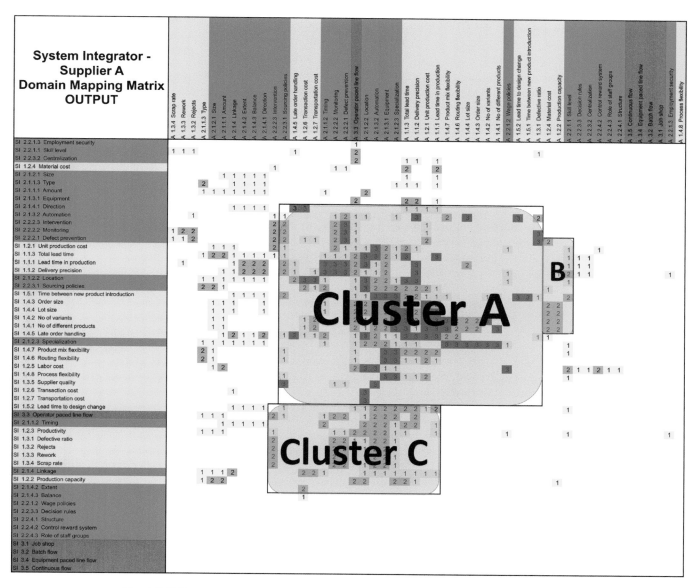

Figure 9.13.5
Detailed view of DMM 4 from figure 9.13.4 after clustering analysis.

for adaptation to customer and business partner needs. Finally, this approach reduces uncertainty in decision making and organizational and process design and enables development of self-organizing networks. In our analysis of linkages between manufacturing strategies and the production system in a collaborative network setting, we could identify three major processes—loops of information processing—to help develop a joint manufacturing strategy for the collaborative network, design the production systems within each company, and thus improve manufacturing and delivery of complete products to the final customer. In other words, the SI and the three suppliers must synchronize their strategies and positions to be more successful, and the MDM pinpoints exactly where this must occur.

The outcome was a self-organized, self-regulated system, where each supplier was proactive and solved planning issues without Kalmar Industries having to act and give orders. This was supported by an open system for information management that all four companies could access.

References

Danilovic, Mike, and Tyson Browning. 2007. Managing Complex Product Development Projects with Design Structure Matrices and Domain Mapping Matrices. *International Journal of Project Management* 25 (3):300–314.

Danilovic, Mike, and Bengt Sandkull. 2005. The Use of Dependence Structure Matrix and Domain Mapping Matrix in Managing Uncertainty in Multiple Project Situations. *International Journal of Project Management* 23:193–203.

Danilovic, Mike, and Mats Winroth. 2006. Corporate Manufacturing Network: From Hierarchy to Self-Organising System. *International Journal of Integrated Supply Management* 2 (1):106–131.

McCormick, William T., Paul J. Schweitzer, and Thomas W. White. 1972. Problem Decomposition and Data Reorganization by a Clustering Technique. *Operations Research* 20 (5):993–1009.

Miltenburg, John. 1995. *How to Formulate and Implement a Winning Plan*. Portland, OR: Productivity Press.

Winroth, Mats, and Mike Danilovic. 2002, April 5–8. *Manufacturing Strategies—Congruence of Manufacturing Processes Within a Supply Chain*. Proceedings of the 13th annual conference of the Production and Operations Management Society, San Francisco, CA.

10 The Future of DSM

The basic DSM methods are several decades old. However, only in recent years has the extension of these methods allowed the application of DSM to such a wide range of problems in engineering management. This book has primarily explained how DSM can be used to develop important insights through modeling, analysis, and scrutiny of DSM models of product, organization, and process architectures—either separately or as multiple domains in linked models. With this book, we have tried to bring new users of DSM up to speed by presenting the fundamental methods and selected applications in each domain of DSM modeling.

Despite several decades of DSM history, we believe that DSM research and application is still at an early stage in its life cycle. Many important lessons have been learned, with more to come. Much research has been done, with more to come. Several DSM software tools are now available, with more to come.

This final chapter of the book reviews some of the broader lessons we have learned about DSM and its application to engineering management. We recognize that the state of the art is continuously evolving through research and development of DSM methods. Finally, we point readers to a number of resources available to support DSM users, including conferences, training, software, and the DSM website. We look forward to the coming decades as the methods yield even greater impact and the field matures.

Lessons Learned So Far

As researchers, teachers, and practitioners, we have used DSM to model, analyze, understand, and improve more than 100 different types of projects in various industrial contexts. Here are some of the most salient lessons we have learned through this experience:

1. **A little modeling goes a long way**   Tremendously important insights can arise through a relatively modest modeling effort.

2. **DSM does not give the answers**   Process owners, system engineers, program managers, and other key stakeholders interpret the DSM models in context. Their greater under-

standing of the situation is facilitated by DSM; it is this understanding that yields the answers which are the keys to improvement.

3. **DSM is not the hammer for every nail**   DSM is part of a larger context of system modeling. Many complementary tools are available and should certainly be used where they better suit the problem at hand.

4. **DSM captures both explicit and implicit knowledge**   A careful modeling effort not only uses available documentation but also reveals the information inside the heads of experts. In this way, DSM may expose underlying assumptions and behaviors that are hidden from traditional views.

5. **Not everyone needs to see the same DSM view**   DSM models may be usefully tailored to provide different perspectives for each stakeholder. DSMs may be summarized as high-level models, expanding areas of specific interest, or annotated to highlight relevant features.

6. **Creating DSM models is not as hard as it may seem**   Many people exposed to DSM for the first time are taken aback by the concise visual representation of the matrix model. They assume that "it takes a PhD from MIT to do it." However, this is absolutely not the case. Most DSM models of the scale we have found to be highly useful can be created in a few weeks of effort by a modeler with access to experts in the domain of the model. Some can be developed much more quickly if the data to build the model are readily available in existing documentation.

7. **All models are wrong; some are useful**   The eminent statistician, George Box, is credited with this statement about statistical models. However, we believe this to be true of DSM models as well. All models make assumptions and simplifications. With experience, you will be able to make the right ones and create useful DSM models that yield important benefits.

8. **DSM facilitates continuous improvement**   DSM modeling can be a highly effective tool in the ongoing improvement efforts for products, organizations, and processes. Much additional benefit can be gained from keeping a DSM as a "living model," where it serves as a repository for organizational learning and is continuously improved along with its subject.

9. **Exploit the visual power of DSM**   In almost every DSM application, we have found that the visual display of DSM can be used to tremendous benefit—perhaps even more powerfully than the analytical tools that support DSM. We have chosen examples in this book to demonstrate many different ways to use colors, graphics, labels, and so forth in DSM models.

10. **DSM experts are valuable and in short supply**   This is by no means intended to be a self-serving statement. We refer to the internal DSM experts that some companies have developed. They are able to provide DSM modeling support to projects throughout the organization.

Research Is Ongoing

We have been engaged in DSM research for much of our professional careers. DSM researchers around the world (many of whom have contributed examples to this book) continue to advance the frontier of DSM knowledge. All of these researchers welcome the involvement of industrial sponsors as field research sites for direct access to real-world problems. In fact, DSM is one of the most directly applied areas of research at the intersection of engineering and management.

The DSM community has been holding an annual conference since 1999. At this event, we have presentations of the latest DSM research from around the world. We also see new DSM applications from industrial practitioners and consultants. Developers of DSM software tools will usually provide demonstrations of the newest capabilities in modeling support. We very much welcome new participants to the conference. We do expect attendees to have a basic familiarity with DSM methods in order to get the most out of the presentations. This book provides that, but tutorials have also been offered before the conference begins for participants who are new to DSM.

At recent conferences, we have seen the realm of DSM modeling expand from engineering into other domains where a network view of interacting entities is helpful. Some of the newer areas where DSM is proving to be useful include social networks, geopolitical problems, healthcare systems, financial systems, and education. This trend is certain to continue in the coming years as more areas recognize the value of a systems perspective on their important issues.

DSM researchers are exploring new data collection and model building methods, which we hope will yield insightful models more easily. We expect DSM researchers to continue the development of better DSM analysis tools for handling large matrices, sequencing and clustering methods for specialized situations, and further innovations in visualization and display of DSM models.

Software and Training

When we began our research in this area, there were no specialized DSM software tools available. Much of our work, therefore, has utilized standard spreadsheet software (e.g., Microsoft Excel) augmented with macros to facilitate manipulation, graphical elements to enhance display, and analysis using either standard functions or additional mathematical software (e.g., MathWorks MATLAB).

Despite the dearth of DSM software solutions, we decided long ago not to get into the software business. Our concerns included losing objectivity and credibility as leading researchers if we were to be personally involved in selling DSM tools. Instead, we have made our research results publicly available through our publications, including new methods and algorithms. Many of these results have been incorporated into the DSM software tools that have emerged in recent years.

Today there are several free and commercial tools for DSM creation, display, and analysis. Some of the software tools will import or export files with Excel or various project management software. We will not review each of the tools here, as the field is still rapidly developing. Although each of the tools has unique strengths and none of the tools does every kind of DSM analysis and display presented in this book, many of them have impressive capabilities today. Indeed the tools are improving every year, and they provide invaluable support for the DSM analyst. All of the tools we are aware of are listed on the DSM website (see below).

There are several sources of DSM training for new practitioners. Courses and tutorials are offered at universities where DSM research is underway and by some consultants specializing in DSM application. Many of these offerings are also listed on the DSM website.

Website

At this time, the DSM website is the primary location where the international DSM community archives publications, posts notices of upcoming conferences, and provides links to training programs, software, and consulting services. The website also contains an online tutorial with basic information about many of the DSM methods presented in this book.

www.dsmweb.org

Index